Christian Barz

Differential Invariants of Prehomogeneous Vector Spaces

Logos Verlag Berlin

λογος

Bibliographic information published by the Deutsche Nationalbibliothek

The Deutsche Nationalbibliothek lists this publication in the Deutsche Nationalbibliografie; detailed bibliographic data is available in the Internet at http://dnb.ddb.de.

ISBN 978-3-8325-4894-0

Logos Verlag Berlin GmbH
Comeniushof, Gubener Str. 47,
D-10243 Berlin
Tel.: +49 (0)30 / 42 85 10 90
Fax: +49 (0)30 / 42 85 10 92
http://www.logos-verlag.de

Differential invariants of prehomogeneous vector spaces

Dissertation

zur Erlangung des akademischen Grades
doctor rerum naturalium
(Dr. rer. nat.)

vorgelegt der Fakultät für Mathematik
der Technischen Universität Chemnitz

von **Dipl.-Math. Christian Barz**

Gutachter :	Prof. Dr. Christian Sevenheck
	Prof. Dr. Claus Hertling
	Prof. Dr. Luis Narváez Macarro
Verteidigung :	05.10.2018

Contents

Introduction

The main topic of this thesis are two types of differential invariants of a divisor $D \subset V = \mathbb{C}^n$. The first one is the classical Bernstein-Sato polynomial, the second one is called spectrum in the main body of the thesis since it is related to the spectrum of isolated hypersurface singularities as studied by Steenbrink, Varchenko and many other people (see e.g. [SS85]). Our main assumption is that D is the discriminant of a regular reductiv prehomogeneous vector space (G, ρ, V), i.e. G is a reductive group, $\rho : G \to \mathrm{Gl}(V)$ is a rational representation such that there is an Zariski open $G-$orbit U in the vector space V and the complement of the open orbit $D = V \backslash U$ is a divisor. We will always assume that the group G is connected. Usually these divisors have complicated non-isolated singularities, so the computation of these invariants is not obvious.

A special case of above mentioned situation is when the divisor is *linear free*, a notion introduced in [BM06]. Linear free divisors are in particular free, that is, their sheaf of logarithmic vector fields $\mathrm{Der}_V(-\log D)$ resp. logarithmic differential forms $\Omega^1_V(\log D)$ is a locally free module over the structure sheaf of the ambient space V. Free divisors were introduced by K. Saito ([Sai80]) and have been studied from many angles in singularity theory. The main class of examples are discriminants in deformation spaces of various objects. Looking at discriminants in quiver representation spaces, one arrives at the more special situation where the coefficients of a basis of logarithmic vector fields, when expressed in the coordinate fields basis, are linear, and in this situation a free divisor is called linear free. In particular, such quiver representations are examples of prehomogenous vector spaces. Notice however that a free divisor is always reduced by definition, which is not the case for arbitrary prehomogenous vector spaces with discriminant divisors. In particular, for non-reduced divisors the sheaves of logarithmic vector fields and logarithmic differential forms are generally not free.

If $D \subset V$ is a linear free divisor, computations of Bernstein polynomials and spectral numbers were done in [GMS09], [GMNRS09], [GS10], [GMS11], [Sev11] and [Sev13]. More precisely, let h be the defining equation of the divisor D, seen as a morphism $h : V \to T := \operatorname{Spec} \mathbb{C}[t]$. Then for a generic linear form $f \in V^*$ we consider the Gauss-Manin-system

$$G(*D) := \frac{h_*\Omega^{n-1}_{V/T}(*D)[\theta, \theta^{-1}]}{(\theta d - df\wedge)h_*\Omega^{n-2}_{V/T}(*D)[\theta, \theta^{-1}]}.$$

It has an abstract interpretation as localized partial Fourier-Laplace transformation of the top cohomology of the direct image complex of the two-component morphism $\Phi = (f, h)$, as is explained below in the summary. The main point in the papers [GMS09, Sev11, Sev13] is that the extension

$$G_0(log\ D) := \frac{h_*\Omega^{n-1}_{V/T}(\log D)[\theta]}{(\theta d - df\wedge)h_*\Omega^{n-2}_{V/T}(\log D)[\theta]}$$

is well defined, has the structure of a free $\mathbb{C}[\theta, t]$-module and specific bases of it can be constructed and studied. Here the freeness of the modules $\Omega^k_{V/T}(\log D)$ (deduced from the fact that $\Omega^1_V(\log D)$ is free for a free divisor) plays a crucial role. Since for general prehomogenous discriminants these modules are no longer free, one problem solved in this thesis is to give a meaningful definition of the Gauss-Manin-system resp. its extension in this more general case.

Spectral numbers encode both the monodromy of the connection on such modules as well Hodge theoretic information given by (integer) Hodge numbers. They can abstractly be defined as a set of numbers (in general complex, but usually rational if the connection has a geometric origin) from the given datum of a free $\mathbb{C}[\theta]$-module with a connection. It is generally hard to calculate these numbers, however, in the situation described above this is possible due a special solution of the so-called Birkhoff problem. The latter consists in finding a special basis of $G_0(\log D)$ such that connection operator takes a particularly simple form. We discuss an algorithm to achieve such a normal form abstractly as well as for some classes of examples.

The other invariant attached to a prehomogenous vector space with a discriminant divisor we are interested in is the (classical) Bernstein-Sato polynomial of a defining equation of the divisor. We recall that for any regular function $h \in \mathcal{O}_V$, the Bernstein-Sato polynomial is the unitary generator $b_h(s) \in \mathbb{C}[s]$ of the ideal of all polynomials $b(s)$ such that a functional equation $P(s)h^{s+1} = b(s)h^s$ holds, where $P \in \mathcal{D}_V[s]$ is a (in general unknown) linear partial differential operator with polynomial coefficients depending on the parameter s. If h is, as in our setup, the defining equation of the discriminant in a prehomogeneous vector space,

then this definition goes back to Sato ([Sat70], [SSM90]). The existence of $b_h(s)$ for general polynomials h is due to Bernstein ([Ber72]) and a similar statement for (germs of) holomorphic functions $h \in \mathcal{O}^{an}_{\mathbb{C}^n,0}$ has been shown by Björk (see [Bjö93]).
If $h\colon (\mathbb{C}^n, 0) \to (\mathbb{C}, 0)$ has an isolated critical point at $0 \in \mathbb{C}^n$, then by [Mal74] the eigenvalues of the monodromy on the cohomology of the Milnor fibre coincide with the exponentials $\exp(2\pi i\alpha)$, where α is a root of $b_h(s)$. In particular the Bernstein-Sato polynomial is a finer invariant of the hypersurface $h^{-1}(0)$ than the monodromy. It is also related to the theory of vanishing cycles of Deligne [Del73] and lead further to the theory of the $V-$filtration of Kashiwara (see [Kas83] and [Mal83]). We remark that there exists generalizations of the Bernstein-Sato polynomial for several polynomials or germs of functions (see [Sab87]) and to an arbitrary variety see [BMS06]) and computer algorithms to compute it (see [ALMM10] for an overview). However a general formula for $b_h(s)$ or $P(s)$ is not known and effective computations are usually difficult to carry out, in particular for functions in many variables and/or (in the algebraic case) of high degree. This is precisely the situation we encounter when studying (defining equations of) prehomogeneous discriminants. Hence a more theoretical approach is needed. It turns out that the spectral numbers mentioned above are very closely related to the roots of the Bernstein-Sato polynomials. As was done in [Sev11] for linear free divisors, we find that they are related by an elementary formula (see the synopsis below), the proof of which, however, needs some $\mathcal{D}$-module theoretic considerations. We use this formula and the computation of spectral numbers to determine the Bernstein-Sato polynomials of a large number of examples.

Summary of the thesis

Let us give a short overview on the different chapters of this thesis. In chapter 1 we give the definition of a prehomogeneous vector space (G, ρ, V) and gather together other basic notions, such as that of a semi invariant, and related facts. Under the assumption that the group G is reductive, we give a first definition of the Bernstein-Sato polynomial $b_h(s)$ for a defining equation h for D. This will become important later, as under this assumption we have a precise description of the differential operator in the functional equation of the Bernstein-Sato polynomial.

In Chapter 2 we introduce the spectrum of meromorphic connection G with a lattice G_0 on the affine line, that is, G is a locally free $\mathcal{O}_{\mathbb{A}^1}(*0)$-module of finite rank with connection ∇ and G_0 is a locally free $\mathcal{O}_{\mathbb{A}^1}$-submodule of same rank. As mentioned above, this definition is motivated by Hodge theoretic considerations for isolated singularities ([SS85]) or polynomial functions ([Sab06]). We also discuss in chapter 2 that if there exists a basis $\underline{\omega}$ of the lattice G_0 solving the Birkhoff problem, i.e. there are constant matrices A_0 and A_∞ such that

$$\nabla_{\theta^2\partial_\theta}\underline{\omega} = \underline{\omega}(A_0 + \theta A_\infty)$$

and if $\underline{\omega}$ is compatible with the Kashiwara-Malgrange $V_\bullet$−filtration (see definitions 2.2.1 and 2.2.3), then the spectrum is given by the eigenvalues of A_∞ (see proposition 2.2.5 for the precise statement).

Chapters 3 and 4 contain the technical core of this work, which allows us to investigate discriminants $D \subset V = \mathbb{A}^n_{\mathbb{C}}$ with a non-reduced defining equation h. Therefore we study in chapter 3 the modules of logarithmic vector fields $\mathrm{Der}(-log\ D)$ and logarithmic k−forms $\Omega^k_V(log\ D)$ in case that the equation h is homogeneous but not necessarily reduced. Moreover we compare the reduced with the non reduced case. As in the sequel we are interested in differential forms relative to the fibres of $h\colon V \to T = \mathrm{Spec}(\mathbb{C}[t])$, which are logarithmic along D, we study the relative logarithmic de Rham complex $(\Omega^\bullet_{V/T}(log\ D), d)$. We obtain a description of the the modules $\Omega^i_{V/T}(log\ D)$ for $i = n-1, n-2$ and this leads to a description of the Jacobian algebra $\frac{\mathcal{O}_V}{df(\mathrm{Der}(-log\ h))}$, which will be an important object in chapter 4, when we study the function f on fibrations, whose central fibre is the discriminant of a regular reductive prehomogeneous vector space. This generalizes the situation of [GMS09], where the central fibre was assumed to be a linear free divisor, and where consequently the modules of logarithmic vector fields $\mathrm{Der}(-log\ D)$ resp. of logarithmic forms $\Omega^1_V(log\ D)$ are free. However, when considering arbitrary defining equations for D this is no longer the case. It turns out that the Lie algebra $\mathfrak{g}$ of G defines a free submodule of $\mathrm{Der}(-log\ D)$, which allows us to prove that if G is reductive and f lies in the dual open orbit $V^*\backslash D^*$, then the Jacobian algebra $\frac{\mathcal{O}_V}{df(\mathrm{Der}(-log\ h))}$ is still a finitely generated free module over $\mathcal{O}_T$.

In chapter 5 we introduce the localized Gauss-Manin system attached to two morphisms $h\colon V \to T = \mathrm{Spec}(\mathbb{C}[t])$ and $f\colon V \to R = \mathrm{Spec}(\mathbb{C}[r])$. More precisely, it is defined as the top cohomology of the direct image complex $\Phi_+\mathcal{O}_V(*D)$ of the sheaf of rational functions along $D = \mathcal{V}(h)$ under the morphism

$$\begin{aligned}\Phi\colon V &\to R \times T\\ v &\mapsto (f(v), h(v)).\end{aligned}$$

The cohomology modules of this complex carry a $\mathcal{D}_{R\times T}$–structure. We determine a representation by relative differential forms and formulas for the ∂_r– and ∂_t–actions. It turns out that after a partial Fourier Laplace transformation the cohomology module of top degree, called *FL-Gauss-Manin system*, is free. Moreover we define a lattice, the *FL-Brieskorn lattice*, and we solve a Birkhoff problem as discussed in chapter 2 for it. More precisely, we obtain the following result:

Theorem. (cf. Theorem 5.2.6) Let (G, V) be a regular reductive prehomogeneous vector space with discriminant D. Let h be an equation of degree $n_h = \deg(h)$ for D (reduced or not) and $f \in V^*\setminus D^*$ be a generic linear form. Then there is a $\mathbb{C}[\theta, t]$–basis $\omega^{(3)} = (\omega_1^{(3)}, \ldots, \omega_{n_h}^{(3)})$ of the FL-Brieskorn lattice $G_0(log\ D)$, such that

$$\nabla\omega^{(3)} = \omega^{(3)} \left[(A_0\theta^{-1} + \theta A_\infty)\frac{d\theta}{\theta} + (-A_0\theta^{-1} - A_\infty + D)\frac{dt}{n_h t} \right]$$

where

$$A_0 = \begin{pmatrix} 0 & 0 & \ldots & 0 & c\cdot t \\ 1 & 0 & \ldots & 0 & 0 \\ \vdots & \vdots & \ldots & \vdots & \vdots \\ 0 & 0 & \ldots & 0 & 0 \\ 0 & 0 & \ldots & 1 & 0 \end{pmatrix}$$

$A_\infty = \operatorname{diag}(\nu_1, \ldots, \nu_{n_h})$ and $D = \operatorname{diag}(0, 1, \ldots, n_h - 1)$ and $(\nu_1, \ldots, \nu_{n_h})$ is the spectrum at infinity of the restriction $G_0(log\ D)_{|t}$.

In chapter 6 we discuss how to obtain the roots of the Bernstein-Sato polynomial of h from the computations of the spectrum of the FL-Brieskorn lattice. This parallels the arguments in [Sev11], but again we do not suppose here that the defining equation h of D is reduced.

We give a definition of the Bernstein-Sato polynomial of the equation h via V–filtration and relate them to the classical one using a functional equation. Furthermore we see that the restriction of the FL-Gauss-Manin system to $\theta \neq 0$ is a hypergeometric $\mathbb{C}[t]\langle\partial_t\rangle$–module. (cf. section 8.1 for the definition) The main result of chapter 6 can be stated as follows:

Corollary. (Corollary 6.2.10) Let h be an equation of the discriminant in a regular reductive prehomogeneous vector space. We denote by $\nu_1, \ldots, \nu_{\deg(h)}$ the spectrum of the FL-Brieskorn lattice and the roots of the Bernstein-Sato polynomial of h by $\alpha_1, \ldots, \alpha_{\deg(h)}$. Then we have

$$\alpha_i = \frac{i - 1 - \nu_i}{\deg(h)} - 1.$$

Chapter 7 is devoted to quivers Q and their representation theory, as they are our main source of examples for regular reductive prehomogeneous vector spaces. We give criteria for regularity and prehomogeneity and explain in this situation a method how to obtain the fundamental semi-invariants of the prehomogeneous vector space from the initial data.

In chapter 8 we present the results of the computation of concrete examples. Moreover we check in these examples the invariant subspace conjecture and the exponent condition (see section 8.1 for the precise definitions and statements.).

The last two chapters 9 and 10 are devoted to two specific series of examples, which are connected by a simple construction that we explain. This construction allows us to produce new classes of examples of prehomogeneous discriminants from known ones.

Danksagung

Ich möchte mich bei all denjenigen bedanken, die mich während der Anfertigung der vorliegenden Dissertation unterstützt haben. Insbesondere bin ich meinem Betreuer Christian Sevenheck für seine uneingeschränkte Unterstützung sehr dankbar. Unsere Zusammenarbeit hat mein mathematisches Wissen deutlich weiterentwickelt. Ebenfalls möchte ich mich bei meinen Arbeitskollegen Alberto, Dmytro und Thomas bedanken.

Chapter 1

Prehomogeneous vector spaces

In the first 3 sections of this chapter we recall basic definitions and facts from the theory of prehomogeneous vector spaces, e.g. the contragredient representation and the Bernstein-Sato polynomials for reductive prehomogeneous vector spaces. We refer for more details of the theory to [SK77], [Gyo91] and [Kim03].

In the first two sections we are in a more general setting and denote by $\mathbb{F}$ an algebraic closed field of characteristic zero. Later in the last two sections we specialize to $\mathbb{F} = \mathbb{C}$.

1.1 Prehomogeneous vector spaces and relative invariants

We denote by G a connected linear algebraic group defined over $\mathbb{F}$ and by $\rho\colon G \to \mathrm{GL}(V)$ a rational $\mathbb{F}-$representation.

Definition 1.1.1. We call a triple (G, ρ, V) a *prehomogeneous vector space*, if there is a Zariski dense $G-$orbit in V, i.e. we have $\overline{\rho(G)v} = V$ for some point $v \in V$.

Remark 1.1.2. If the representation is clear from the context, we simplify the notation and write (G, V) instead of (G, ρ, V).
In particular in our applications G will usually be a subgroup of $\mathrm{GL}(V)$. We notate the group action by $g.v := \rho(g)v$ and use this notation for other actions, when the action is clear from the context, e.g. we write $A.v := d\rho(A)v$ when considering the action of the Lie algebra $\mathfrak{g}$ of G given by the differential $d\rho$.

We recall the *orbit-lemma* from the theory of algebraic groups:

Lemma 1.1.3. ([Hum75] II.8.3 Proposition) Let G be an algebraic group acting on an algebraic variety X, then each orbit is a smooth and locally closed subset of X, i.e. an orbit is open in its closure. The boundary of an orbit is a union of orbits of strictly lower dimension.

Hence by the orbit-lemma a dense orbit in a prehomogeneous vector space (G, ρ, V) is open and unique, as V is irreducible.
Indeed if $O_1, O_2 \subset V$ are two dense orbits, then they are open and their intersection is non empty by the irreducibility of V. Since for any $v \in O_1 \cap O_2$ we have $G.v = O_1$ and $G.v = O_2$, hence a dense orbit is unique.

Proposition 1.1.4. ([Kim03] Proposition 2.2) For each point $v \in V$ of a triple (G, ρ, V) the following conditions are equivalent :

1. $\overline{\rho(G)v} = V$,
2. $\rho(G)v$ is a Zariski open subset of V,
3. $\dim(G_v) = \dim(G) - \dim(V)$, where $G_v = \{g \in G \mid \rho(g)v = v\}$,
4. $\dim(\mathrm{Lie}(G_v)) = \dim(\mathrm{Lie}(G)) - \dim(V)$, where the Lie algebra of G_v is given by $\mathrm{Lie}(G_v) = \{A \in \mathrm{Lie}(G) \mid d\rho(A)v = 0\}$ and
5. $\{d\rho(A)v \mid A \in \mathrm{Lie}(G)\} = V$.

Definition 1.1.5. Let (G, ρ, V) be a prehomogeneous vector space. We call the complement S of the open $G-$orbit the *singular set*. The points of $V \backslash S$ are called *generic points*. If the complement of the open orbit is a divisor, then we call it *the discriminant* and denote it by D.

We recall that a rational morphism of algebraic groups $\chi \colon G \to \mathbb{G}_m = \mathrm{GL}_1(\mathbb{F})$ is called a *rational character*, e.g. $\det \colon \mathrm{GL}_n(\mathbb{F}) \to \mathbb{F}^*$ is a rational character. We denote by $X(G) = \mathrm{Mor}(G, \mathrm{GL}_1(\mathbb{F}))$ the group of rational characters of G.

Definition 1.1.6. Let (G, V, ρ) be a prehomogeneous vector space. A non-zero rational function f on V is called *semi-invariant* of weight χ, if there exists a rational character $\chi \in X(G)$, such that

$$f(\rho(g)x) = \chi(g)f(x)$$

holds for all $x \in V \backslash S$ and $g \in G$.
A semi-invariant f is called *invariant*, if the corresponding character is trivial, i.e. we have $f(\rho(g)x) = f$ for all $g \in G$.

Remark 1.1.7. When studying a $G-$variety V one often considers $G-$invariant functions on V. But in the case of a prehomogeneous vector space (G, ρ, V) a $G-$invariant function f is constant on an open and dense subset and hence f is constant on V. Conversely by [Kim03] proposition 2.4 if any $G-$invariant function of a triple (G, V, ρ) is a constant, then the triple (G, V, ρ) is a prehomogeneous vector space.
Hence semi-invariants play a basic role in the theory of prehomogeneous vector spaces.
We collect basic properties of semi-invariants in the following lemma:

Lemma 1.1.8. ([SSM90] §1 Proposition 2)

1. If two semi-invariants f_1 and f_2 have the same character, then f_1 is a constant multiple of f_2, i.e. there is a $c \in \mathbb{F}$ such that $f_1 = cf_2$.
2. Any prime divisor of a semi-invariant is a semi-invariant.
3. Semi-invariants are homogeneous.
4. Semi-invariants corresponding to multiplicatively independent characters are algebraic independent.

Let (G, ρ, V) a prehomogeneous vector space with discriminant D. By the orbit lemma the boundary of an orbit consists of $G-$orbits of lower dimension. Therefore the discriminant D of a prehomogeneous vector space (G, ρ, V) is preserved by G and an equation for D is a semi-invariant.

Lemma 1.1.9. The zero locus of every non-zero semi-invariant f is contained in D.

Proof. Let f be a semi-invariant. Suppose $x \in V \backslash D$ and $f(x) = 0$, then we have $f(g.x) = \chi(g)f(x) = 0$ for all $g \in G$ and f vanishes on a dense open subset of V. Hence f vanishes on all of V. □

Proposition 1.1.10. ([SK77] §4 Proposition 5) Let (G, ρ, V) be a prehomogeneous vector space with singular set S and let $S_i = \mathcal{V}(f_i) \subset S$,$i = 1, \dots, r$, be the irreducible components of codimension one. Then each S_i is the zero locus of some irreducible polynomial f_i, i.e. for $i = 1, \dots, r$ we have

$$S_i = \{v \in V | f_i(v) = 0\}.$$

The irreducible polynomials $f_1, \dots, f_r$ are algebraically independent semi-invariants and every semi-invariant f can be uniquely written as

$$f = cf_1^{e_1} \dots f_r^{e_r}$$

for $c \in \mathbb{F}^*$ and $(e_1, \dots, e_r) \in \mathbb{Z}^r$.

Definition-Lemma 1.1.11. The semi-invariants $f_1, \ldots, f_r$ of proposition 1.1.10 are called *fundamental semi-invariants.*
They correspond to linear independent characters and generate the ring spanned by the semi-invariants. In particular there are non trivial semi-invariant if and only if the singular set S has an irreducible component of codimension one.

Remark 1.1.12. We remark that in the literature semi-invariants are also called relative invariants and the fundamental semi-invariants are called basic relative invariants.

1.2 The contragredient representation

For a triple (G, ρ, V) we denote by $V^* := \{f : V \to \mathbb{F} |\ f \text{ is a linear map}\}$ the dual space of V and write $\langle \cdot, \cdot \rangle$ for the natural pairing between V and V^*, i.e. $\langle v, f \rangle := f(v)$ for $v \in V$ and $f \in V^*$.
The relation

$$\langle \rho(g)v, \rho^*(g)f \rangle = \langle v, f \rangle$$

defines a rational representation

$$\rho^* : G \to \mathrm{GL}(V^*),$$

which is called *the contragredient representation* of ρ and the triple (G, ρ^*, V^*) is called the *dual triple* of (G, ρ, V).

We will describe the contragredient representation ρ^* more explicit. Therefore we choose a basis $v_1, \ldots, v_n$ of V and a dual basis $v_1^*, \ldots, v_n^*$ of V^*, i.e. we have $\langle v_i, v_j^* \rangle = \delta_{ij}$, where δ_{ij} denotes the Kronecker delta. We identify the elements of V and V^* with $\mathbb{F}^n$ by

$$v = \sum x_i v_i \leftrightarrow x = (x_1, \ldots, x_n)^T$$

and

$$v^* = \sum y_i v_i^* \leftrightarrow y = (y_1, \ldots, y_n)^T.$$

Hence the natural pairing is given by

$$\langle v, v^* \rangle = \langle \sum x_i v_i, \sum y_j v_j^* \rangle = x^T y.$$

Furthermore we denote the matrices of $\rho(g)$ and $\rho^*(g)$ by the same symbols, thus the defining relation becomes

$$x^T \rho^T(g) \rho^*(g) y = \langle \rho(g)v, \rho^*(g)v^* \rangle = \langle v, v^* \rangle = x^T y.$$

As this has to hold for all x, y, we conclude $\rho^*(g) = \rho^{-1}(g)^T$ for all $g \in G$.

In general the dual triple of a prehomogeneous vector space is not prehomogeneous as the following example shows:

Example 1.2.1. ([Kim03] Remark 2.1) Let $V = \mathbb{C}^2$ and

$$G = \{\begin{pmatrix} 1 & b \\ 0 & a \end{pmatrix} \mid a \neq 0\}.$$

We consider the $G-$action given by

$$\rho(g)\begin{pmatrix} x_1 \\ x_2 \end{pmatrix} = \begin{pmatrix} 1 & b \\ 0 & a \end{pmatrix}\begin{pmatrix} x_1 \\ x_2 \end{pmatrix} = \begin{pmatrix} x_1 + bx_2 \\ ax_2 \end{pmatrix}.$$

It is easy to see that (G, ρ, V) is a prehomogeneous vector space with singular set $S = \{\begin{pmatrix} x_1 \\ 0 \end{pmatrix} \in \mathbb{C}^2\}$. In this case the contragredient action is given by

$$\rho^*(g)\begin{pmatrix} y_1 \\ y_2 \end{pmatrix} = \begin{pmatrix} 1 & 0 \\ -\frac{b}{a} & \frac{1}{a} \end{pmatrix}\begin{pmatrix} y_1 \\ y_2 \end{pmatrix} = \begin{pmatrix} y_1 \\ -\frac{b}{a}y_1 + \frac{1}{a}y_2 \end{pmatrix}.$$

Since $f(y_1, y_2) = y_1$ is a non constant absolute invariant, (G, ρ^*, V^*) can not be a prehomogeneous vector space by remark 1.1.7.

As for our application it is necessary that the dual triple of a prehomogenous vector space is prehomogeneous, we briefly recall a condition when the prehomogeneity of (G, V, ρ) imply the prehomogeneity of (G, ρ^*, V^*). We refer for the details of the dicussion to [SSM90] §2 and [Kim03] §2.

Let f be a semi-invariant and S the singular set of a prehomogeneous vector space (G, ρ, V). We define a mapping $\phi_f \colon V \backslash S \to V^*$ by

$$\phi_f(x) = \frac{1}{f(x)} \sum_{i=1}^{n} \frac{\partial f}{\partial_{x_i}}(x) v_i^* = \frac{1}{f(x)} \begin{pmatrix} \frac{\partial f}{\partial_{x_1}(x)} \\ \vdots \\ \frac{\partial f}{\partial_{x_n}}(x) \end{pmatrix}.$$

The map ϕ_f does not depend on the choice of basis and is $G-$invariant by [Kim03] proposition 2.13, i.e. $\phi_f(\rho(g)x) = \rho^*\phi_f(x)$. So the image of ϕ_f is a $G-$orbit. Indeed let $v \in V \backslash S$, then we have

$$\phi_f(V \backslash S) = \phi_f(\rho(G)v) = \rho^*(G)\phi_f(v) \subset V^*.$$

If this orbit is Zariski dense in V^*, then the semi-invariant is called *non-degenerate*. We call the prehomogeneous vector space (G, ρ, V) *regular*, if there is a nondegenerated semi-invariant. Thus by definition the dual of a regular prehomogeneous vector space is a prehomogeneous vector space.

Theorem 1.2.2. ([Kim03] Theorem 2.16) If (G, ρ, V) is a regular prehomogeneous vector space, then the dual triple (G, ρ^*, V^*) is a regular prehomogeneous vector space.
Moreover $\phi_f \colon V\backslash S \to V^*\backslash S^*$ is a biregular isomorphism, where S (resp. S^*) denotes the singular set in V (resp. V^*).

Remark 1.2.3. Although we will not use it, we remark that if a semi-invariant f satisfies $\det(\frac{\partial^2 log\ f}{\partial_{x_i}\partial_{x_j}}) \not\equiv 0$, then f is nondegenerated.

1.3 Reductive prehomogeneous vector spaces and the Bernstein-Sato polynomial

For our applications we are interested in prehomogeneous vector spaces (G, ρ, V), such that the singular set is a hypersurface and the dual triple (G, ρ^*, V^*) is a prehomogeneous vector space.
We have seen in the last section that regularity of (G, ρ, V) implies prehomogeneity of (G, ρ^*, V^*). Moreover by theorem 1.2.2 if the singular set S of (G, ρ, V) is a hypersurface, then the singular set S^* of (G, ρ^*, V^*) is a hypersurface. We will see that both conditions coincide, when (G, ρ, V) is a reductive prehomogeneous vector space over $\mathbb{C}$.

Definition 1.3.1. We call a prehomogeneous vector space (G, ρ, V) a *reductive prehomogeneous vector space*, if G is a reductive algebraic group.

We now specialize to the case $\mathbb{F} = \mathbb{C}$. We denote by $\overline{}$ the complex conjugate and by U^c the Zariski closure of a subset $U \subset V$. We identify V and V^* with $\mathbb{C}^n$ by choosing a basis and its dual basis respectively. Moreover we assume G to be reductive.

Theorem 1.3.2. ([Kim03] Theorem 2.28) For a reductive prehomogeneous vector space (G, ρ, V) over $\mathbb{C}$ the following statements are equivalent :

1. (G, ρ, V) is a regular prehomogeneous vector space.

2. The singular set S is a hypersurface.

3. The orbit $\rho(G)v = V\backslash S$ is an affine variety.

4. Each generic isotropy group G_v, $v \in V \backslash S$, is reductive.

Another crucial point, why we are considering reductive prehomogeneous vector spaces, is that we have a precise description of the Bernstein-Sato polynomial[1] $b_h(s)$ for a semi-invariant f. This is due to the existence of an unitary basis, as we will now see following [Kim03].

Because G is reductive (over $\mathbb{C}$), G contains a compact Zariski dense subgroup K (see for example [OV12] chapter 5). Since any compact subgroup of $\mathrm{GL}_n(\mathbb{C})$ is conjugated to a subgroup of the unitary group

$$U_n = \{g \in \mathrm{GL}_n(\mathbb{C}) \mid \overline{g}^T g = \mathbf{1}\},$$

we may assume $\rho(K) \subset U_n$. Therefore the contragredient action of K is given by

$$\rho^*(g) = \rho^{-1}(g)^T = \overline{\rho(g)},\ \forall g \in K.$$

Let $f(x) \in \mathcal{O}_V$ be a semi-invariant polynomial of degree $\deg(f) = r$ with associated character χ. We set $f^*(y) := \overline{f(\overline{y})} \in \mathcal{O}_{V^*}$ and obtain:

$$f^*(\rho^*(g)y) = f^*(\overline{\rho(g)})y) = \overline{f(\rho(g)\overline{y})} = \overline{\chi(g)} \cdot \overline{f(\overline{y})} = \overline{\chi(g)} f^*(y)$$

for all $g \in K$. So f^* is a semi-invariant for K.
We claim that it is also a semi-invariant for G. Because $|\chi(K)|$ is a compact subgroup of $\mathbb{R}_{>0}$, we have $|\chi(g)| = 1$ for all $g \in K$. We conclude that $\overline{\chi(g)} = \chi(g)^{-1}$ holds for all $g \in K$. Hence K is contained in the Zariski closed subset

$$\{g \in G \mid f^*(\rho^*(g)y) = \chi^{-1}(g) f^*(y)\}$$

of G. Because K is Zariski dense in G, we conclude that

$$f^*(\rho^*(g)y) = \chi^{-1}(g) f^*(y)$$

holds for all $g \in G$ and $y \in V^*$. Hence f^* is a semi-invariant with associated character χ^{-1}.
Next we introduce the the Bernstein-Sato polynomial. For

$$f^*(x) = \sum_{i_1 + \ldots + i_n = r} f_{i_1 \ldots i_n} x_1^{i_1} \ldots x_n^{i_n}$$

we put

$$f^*(D_x) = \sum_{i_1 + \ldots + i_n = r} f_{i_1 \ldots i_n} \partial_{x_1}^{i_1} \ldots \partial_{x_n}^{i_n}.$$

[1]The Bernstein-Sato polynomial is also called the b–function.

This is a differential operator with constant coefficient satisfying

$$f^*(D_x)e^{\langle x,y\rangle} = f^*(y)e^{\langle x,y\rangle},$$

where $\langle x, y\rangle = x_1y_1 + \ldots x_ny_n$. For $g \in \mathrm{GL}_n(\mathbb{C})$ the equality

$$\begin{pmatrix} x'_1 \\ \vdots \\ x'_n \end{pmatrix} = g. \begin{pmatrix} x_1 \\ \vdots \\ x_n \end{pmatrix}$$

implies

$$\begin{pmatrix} \partial_{x'_1} \\ \vdots \\ \partial_{x'_n} \end{pmatrix} = (g^{-1})^T \begin{pmatrix} \partial_{x_1} \\ \vdots \\ \partial_{x_n} \end{pmatrix}$$

and we conclude

$$f^*(D_{\rho(g)x}) = f^*(\rho^*(g)D_x) = \chi^{-1}(g)f^*(D_x).$$

Now let s be an extra parameter. We put $\phi(x) = f^*(D_x)f(x)^{s+1}$ and because of $f(\rho(g)x)^{s+1} = \chi^{s+1}(g)f^*(x)^{s+1}$ we have :

$$\phi(\rho(g)x) = \chi^s(g)\phi(x).$$

Moreover because of $f(\rho(g)x)^s = \chi^s(g)f(x)^s$ we have, that $f(x)^s$ and $\phi(x)$ are two semi-invariants, which have the same character. Hence by lemma 1.1.8 one is a constant multiple of the other. In other words the ratio $\frac{f^*(D_x)f(x)^{s+1}}{f(x)^s}$ does not depend on $x \in V\backslash S$, but depend on s. So we define

$$b(s) := \frac{f^*(D_x)f(x)^{s+1}}{f(x)^s}$$

and obtain

$$f^*(D_x)f(x)^{s+1} = b(s)f(x)^s. \tag{1.1}$$

We claim that $b(s) = b_0s^d + b_1s^{d-1} + \ldots + b_r$ with $b_0 \neq 0$, i.e. $b(s)$ is a polynomial of degree $r = \deg(f)$. Indeed since we have

$$\begin{aligned}
\frac{\partial}{\partial_{x_i}} f(x)^{s+1} &= (s+1)f(x)^s \frac{\partial f}{\partial_{x_i}}(x) \\
\frac{\partial^2}{\partial_{x_i}\partial_{x_j}} f(x)^{s+1} &= s(s+1)f(x)^{s-1} \frac{\partial f}{\partial_{x_i}} \frac{\partial f}{\partial_{x_j}}(x) \\
&+ (s+1)f(x)^s \frac{\partial^2 f}{\partial_{x_i}\partial_{x_j}}(x),
\end{aligned}$$

$b(s)$ is a polynomial in s satisfying $\deg b(s) \leq \deg f^* = r$. Moreover we can assume by an unitary base change, that $f(1, 0, \ldots, 0) \neq 0$, i.e. the coefficient $f_{r,0,\ldots,0}$ of f is not zero. If the degree of $b(s)$ is $d \leq r$, then there exists a constant c, such that $|b(s)| \leq c(s+1)^d$ holds for all $s \geq 0$. For any $m \in \mathbb{N}$ the functional equation (1.1) implies

$$f^*(D_x)^m f(x)^m = b(0)b(1)\ldots b(m-1),$$

and by a direct calculation we get

$$f^*(D_x)f(x) = \sum_{i_1+\ldots+i_n=r} |f_{i_1\ldots i_n}|^2 (i_1)!\ldots(i_n)!$$

and conclude

$$|f_{r,0,\ldots,0}|^{2m}(mr)! \leq f^*(D_x)^m f(x)^m = b(0)b(1)\ldots b(m-1) \leq c(m!)^r.$$

Because of $\frac{(mr)!}{(m!)^r} \geq 1$, we have $(m!)^{r-d} \leq (\frac{c}{|f_{d,0,\ldots,0}|^2})^r$ and therefore $r = d$. We summarize the discussion :

Proposition 1.3.3. ([Kim03] Proposition 2.22) Let (G, ρ, V) be a reductive prehomogeneous vector space over $\mathbb{C}$ and let $f(x)$ be a semi-invariant of degree r. Then there is a polynomial $b(s)$ of degree r with

$$f^*(D_x)f(x)^{s+1} = b(s)f(x)^s.$$

Definition 1.3.4. We call the polynomial $b(s)$ defined by the functional equation (1.1) the $b-$*function* of the semi-invariant f.

We remark that the $b-$function of a semi-invariant f of a reductive prehomogeneous vector space is (up to a constant multiple) the Bernstein-Sato polynomial of f, which we will define in chapter 6. Moreover we note that reductive prehomogeneous vector spaces were studied in [Gyo91] without the regularity condition.

1.4 The differential of the orbit map

We will now study a special situation, which will be useful, when studying restrictions of a linear function f to the fibres of a morphism $h\colon V = \mathbb{A}^n_{\mathbb{C}} \to \mathbb{C}$ in chapter 4. We start in the following setting:
Let $D \subset V$ be a divisor defined by a homogeneous polynomial h. We denote by $x_1, \ldots, x_n$ coordinates on V and by $\partial_1, \ldots, \partial_n$ the corresponding

partial derivatives and set $\deg(x_i) = -\deg(\partial_i) = 1$.
We consider the group

$$G := G_D := \{g \in \mathrm{GL}(V) \mid g(D) \subset D\}^0,$$

which is the connected component of the identity of the isotropy group of D. Clearly G is a linear algebraic group. We denote by $\mathfrak{g}$ its Lie algebra and by

$$L := L_D = \{x^T A \partial = \sum_{ij} x_i a_{ij} \partial_j \mid x^T A\partial(h) \in \mathbb{C}h\}$$

the Lie subalgebra of $\mathrm{Der}_V(-log\ D) := \{\xi \in \mathrm{Der}_V \mid \xi.h \subset (h)\}$, which consists of vector fields of weight zero. By [GMNRS09] lemma 2.2 we have:

Lemma 1.4.1. The map

$$\begin{aligned} \phi\colon \quad \mathfrak{g} &\to L_D \\ A &\mapsto x^T A^T \partial \end{aligned}$$

is an Lie algebra isomorphism.

We conclude from lemma 1.4.1 that the induced map

$$\phi\colon \mathcal{O}_V \otimes \mathfrak{g} \to \mathrm{Der}_V(-log\ D)$$

is injective. This observation will be a key information in chapter 4. We remark that by [GMNRS09] lemma 2.2-2.4 the induced map ϕ is an isomorphism if and only if D is a *linear free divisor.*

Remark 1.4.2. We recall that a divisor $D \subset V$ is called *free*, if the sheaf $\mathrm{Der}(-log\ D)$ is locally free. If there is moreover a global basis $\xi_1, \ldots, \xi_n$, such that $\xi_i(x_j) \in \mathbb{C}[V]_1 \cup \{0\}$ holds for all i, j, then the divisor D is called a *linear free divisor.*
On the other hand it is well known (e.g. [GS10] §2) that a linear free divisor is special case of the discriminant of prehomogeneous vector space (G, ρ, V) characterized by the conditions:

1. $G \subset \mathrm{GL}(V)$ is connected and $\dim(G) = \dim(V)$,
2. $\rho = \mathrm{id}\colon \mathrm{G} \to \mathrm{GL}(V)$ and
3. $h(x) := \det(d\rho(A_1)x| \ldots |d\rho(A_n)x)$ is a reduced equation for D, where $A_1, \ldots, A_n$ is a basis of $\mathfrak{g}$.

In this thesis we will neither require that h is reduced nor is the semi-invariant defined in remark 1.4.2 (3).[2] But we will assume that (G, ρ, V) is a regular reductive prehomogeneous vector space with discriminant D.

Because we study linear functions on the fibration of h in chapter 4, we explain next how the objects are connected.

Let f be a homogeneous polynomial of degree one, which we consider as an element of the dual space V^*. We denote by

$$\begin{array}{rcl} \alpha_f \colon G & \to & V^* \\ g & \mapsto & \rho^*(g)f \end{array}$$

the orbit map (at f) and by $d\alpha_f$ its differential. To study the differential explicitly, we choose a basis of V (and dually of V^*) and consider G as a subgroup of $\mathrm{GL}_n(\mathbb{C})$. Then the differential of the orbit map (at the identity) is

$$\begin{array}{rcl} d\alpha_f \colon \mathfrak{g} & \to & T_f(G.f) \\ A & \mapsto & -A^T f. \end{array}$$

We look at the following diagram

$$\begin{array}{ccc} L_D & \xrightarrow{df} & \mathfrak{m}\,, \\ \uparrow{\scriptstyle\phi} & \nearrow_{d\alpha_f} & \\ \mathfrak{g} & & \end{array}$$

where $\mathfrak{m}$ denotes the maximal ideal of $\mathcal{O}_V$ at zero and the map df is defined as the restriction of

$$\begin{array}{rcl} \mathrm{Der}(-log\ D) & \to & \mathcal{O}_V \\ \xi & \mapsto & df(\xi) = \xi(f). \end{array}$$

to L_D. In particular because L_D consists of vector fields of weight zero and f is homogeneous of degree one, the image of the restriction lies in $\mathfrak{m}$.

Because it will be important in chapter 4, when studying the Jacobian algebra

$$\frac{\mathcal{O}_V}{df(\mathrm{Der}(-log\ h)},$$

we do the following explicit calculation :

[2] Although in examples we will often study D by the equation $h = \det(A_1.x|\dots|A_n.x)$.

For $f = \sum_l f_l x_l \leftrightarrow (f_1, \ldots, f_n)^T \in V^*$, $A = (a_{ij}) \in \mathfrak{g}$ we have :

$$\begin{aligned} d\alpha_f(A) &= (-\sum_j a_{ji} f_j)_{i=1,\ldots,n} \\ df \circ \phi(A) &= df(x^T A^T \partial) \\ &= (\sum_{i,j} x_i a_{ji} \partial_j)(\sum_l f_l x_l) \\ &= (\sum_j f_j a_{ji})_{i=1,\ldots,n}. \end{aligned}$$

So if we put $\tilde{\mathfrak{g}} := \phi(\mathfrak{g}) \subset \mathrm{Der}(-log\ D)$, the ideals $\mathcal{O}_V df(\tilde{\mathfrak{g}})$ and $\mathcal{O}_V \alpha_f(\mathfrak{g})$ are equal.

Corollary 1.4.3. For $f \in V^*$ the differential of the orbit map factors through

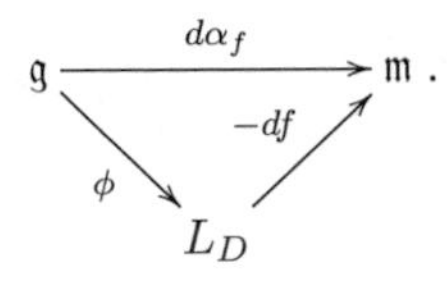

Chapter 2

The spectrum and the Birkhoff problem

In this chapter we introduce the notion of the spectrum of a meromorphic connection with a lattice, where we follow in the presentation [Sab06]. We introduce the notion of a V^+–solution, which is a special form of a Birkhoff problem. We will see that in the case of a V^+–solution the spectrum is given by the eigenvalues of a special matrix.

2.1 The spectrum of a meromorphic connection with a lattice

Let $U_0 = \mathrm{Spec}(\mathbb{C}[\tau])$ and $U_\infty = \mathrm{Spec}(\mathbb{C}[\theta])$ denote the two standard charts of $\mathbb{P}^1$, where $\theta = \tau^{-1}$ on $U_0 \cap U_\infty$.
We denote by $0 = \{\tau = 0\}$ and by $\infty = \{\theta = 0\}$. Let G be a free $\mathbb{C}[\tau, \tau^{-1}]$–module of finite rank μ, equipped with a derivation ∂_τ giving G the structure of a left $\mathbb{C}[\tau, \tau^{-1}]\langle\partial_\tau\rangle$–module. Following [Sab06] Part I we call such a module G a *meromorphic connection*, as it corresponds to rational bundle with connection with singularities only at $0 = \{\tau = 0\}$ and $\infty = \{\theta = 0\}$.
We recall that a *lattice at* $0 = \{\tau = 0\}$ *(resp. at* $\infty = \{\theta = 0\}$) of G is a free $\mathbb{C}[\tau]$– (resp. $\mathbb{C}[\theta]$–) submodule $\tilde{G}$, such that $G = \mathbb{C}[\tau, \tau^{-1}] \otimes_{\mathbb{C}[\tau]} \tilde{G}$ (resp. $G = \mathbb{C}[\theta, \theta^{-1}] \otimes_{\mathbb{C}[\theta]} \tilde{G}$). It is known that a lattice has rank μ.
We assume that the singularity at $0 = \{\tau = 0\}$ is *regular*, i.e. there exists a lattice at $0 = \{\tau = 0\}$ stable under the operator $\tau\partial_\tau$. The singularity at $\infty = \{\theta = 0\}$ may be irregular, but we will assume there is a lattice G_0 of *type one* at infinity, i.e. the lattice G_0 is stable under the operator $\theta^2\partial_\theta$.

The lattice G_0 defines an increasing filtration on G by putting

$$G_k := \theta^{-k} G_0.$$

On the other hand there is the Kashiwara-Malgrange filtration, whose definition we briefly recall in this situation[1]
Let $V_\bullet \mathbb{C}[\tau]\langle\partial_\tau\rangle$ be the increasing filtration of $\mathbb{C}[\tau]\langle\partial_\tau\rangle$ defined by

$$\begin{aligned} V_{-k}\mathbb{C}[\tau]\langle\partial_\tau\rangle &= \tau^k \mathbb{C}[\tau]\langle\tau\partial_\tau\rangle \text{ for } k \geq 0 \\ V_k\mathbb{C}[\tau]\langle\partial_\tau\rangle &= V_{k-1}\mathbb{C}[\tau]\langle\partial_\tau\rangle + \partial_\tau V_{k-1}\mathbb{C}[\tau]\langle\partial_\tau\rangle \text{ for } k \geq 1. \end{aligned}$$

The *Kashiwara-Malgrange filtration* (or *$V-$filtration*) is the unique increasing exhaustive filtration $V_\bullet G$ of G, indexed by $\mathbb{Q}$, satisfying the following properties for all $\alpha \in \mathbb{Q}$:

1. $V_\alpha G$ is a free $\mathbb{C}[\tau]-$module and $\mathbb{C}[\tau,\tau^{-1}] \otimes_{\mathbb{C}[\tau]} V_\alpha = G$,
2. $\tau V_\alpha G \subset V_{\alpha-1} G$,
3. $\partial_\tau V_\alpha G \subset V_{\alpha+1} G$ and
4. the action $\tau\partial_\tau + \alpha$ is nilpotent on $gr_\alpha^V G = V_\alpha G / V_{<\alpha} G$.

Moreover there exists a finite subset $\Lambda \subset [0,1)$ such that $gr_\alpha^V G = 0$ for all $\alpha \notin \Lambda + \mathbb{Z}$.

Remark 2.1.1. The Kashiwara-Malgrange filtration exists for an holonomic $\mathcal{D}-$module. Indeed for an algebraic variety X a $\mathcal{D}_X-$module M is called *holonomic*, if the dimension of the characteristic variety is equal to the dimension of X. By [HT07] prop. 2.2.5 the restriction $G_{|U_0}$ is holonomic and we conclude using [MM04] proposition 4.3.5 and 4.4-2 that the Kashiwara-Malgrange filtration exists in our situation.
However we will see in the more special situation of the next section that the Kashiwara-Malgrange filtration will be given concretely by a special solution of a Birkhoff problem (cf. proposition 2.2.5 and its proof).
We refer to [Sab06] § 1, [MM04] 4 and [DS03]§ A.b.3 for detailed discussion of the Kashiwara-Malgrange filtration.

Lemma 2.1.2. ([Sab06] 1.3 p.179) For every $\alpha \in \mathbb{Q}$ the $\mathbb{C}[\tau]-$module $V_\alpha G$ is free of rank μ and defines a logarithmic connection with pole at 0 on U_0.

[1] We refer to section 6.1 for a more detailed discussion of the $V-$filtration.

Definition 2.1.3. For $\alpha \in \mathbb{Q}$ we set

$$\nu_\alpha = \dim(\frac{V_\alpha G \cap G_0}{V_\alpha G \cap G_{-1} + V_{<\alpha} G \cap G_0})$$

and call the set of pairs (α, ν_α) with $\nu_\alpha \neq 0$ the *spectrum* $SP(G, G_0)$ of (G, G_0) and the polynomial

$$\mathrm{SP}(G, G_0, s) := \prod_{\alpha \in \mathbb{Q}} (s + \alpha)^{\nu_\alpha}.$$

the *spectral polynomial* of (G, G_0).

Lemma 2.1.4. ([Sab06] lemma 1.6) The degree of the spectral polynomial $\mathrm{SP}(G, G_0, s)$ is equal to μ.

2.2 The Birkhoff problem

Let G be a meromorphic connection having poles only at $0 = \{\tau = 0\}$ and $\infty = \{\theta = 0\}$ and let G_0 be a lattice at $\infty = \{\theta = 0\}$.

Definition 2.2.1. A basis $\omega \subset G_0$ is called *a solution to the Birkhoff problem*, if in the basis ω the connection $\nabla_{\theta^2 \partial_\theta}$ is given by

$$\nabla_{\theta^2 \partial_\theta} \omega = \omega(A_0 + \theta A_\infty),$$

where A_0 and A_∞ are constant $\mu \times \mu-$matrices.
Following [Sai89] §3 we call such a basis ω a *good basis*, if the spectrum of A_∞ is equal to the spectrum $\mathrm{SP}(G, G_0)$.
Obviously if $\omega = (\omega_1, \ldots, \omega_\mu)$ is a solution to the Birkhoff problem, then $\bigoplus_{i=1}^{\mu} \mathbb{C}[\theta]\omega_i$ is a lattice of type 1 at $\infty = \{\theta = 0\}$.
On the other hand if we put $G^0 = \bigoplus_{i=1}^{\mu} \mathbb{C}[\tau]\omega_i$ and pull back the connection ∇_{∂_θ} along $\tau = \tau^{-1} = \theta$. Then we have $-\nabla_{\partial_\tau} = \nabla_{\theta^2 \partial_\theta}$ and obtain

$$\nabla_{\tau \partial_\tau} \omega = -\omega(A_0 + \tau^{-1} A_\infty)$$

and hence ∇_{∂_τ} has a regular singularity at $0 = \{\tau = 0\}$.
By the following proposition also the converse holds:

Proposition 2.2.2. ([Sab06] 4.1 or [Sab07] p.159ff) The following statements are equivalent:

1. A basis ω of G_0 is a solution to the Birkhoff problem.
2. There is a lattice $G^0 \subset G$ at $0 = \{\tau = 0\}$, which is stable under $\tau\partial_\tau$, such that
$$(G^0 \cap G_0) \oplus G_{-1} = G_0.$$

In order to relate a solution of the Birkhoff problem to the spectrum $\mathrm{SP}(G, G_0)$, we have to consider solutions of the Birkhoff problem, which are compatible with the $V-$filtration, cf. [DS03] Appendix B.

Definition 2.2.3. Let $\omega \subset G_0$ be a solution to the Birkhoff problem and $G^0 = \bigoplus_{i=1}^{\mu} \mathbb{C}[\tau]\omega_i$. We say ω is *compatible* with the filtration $V_\bullet G$ (or ω is a $V-$*solution* for short), if for all $\alpha \in \mathbb{Q}$ holds
$$G_0 \cap V_\alpha G = (G_0 \cap G^0 \cap V_\alpha) \oplus (\theta G_0 \cap V_\alpha).$$
We call such a basis ω a V^+-*solution* if moreover
$$(\tau\partial_\tau + \alpha)(G_0 \cap G^0 \cap V_\alpha G) = (G_0 \cap G^0 \cap V_{<\alpha} G) \oplus \tau(G_0 \cap G^0 \cap V_{\alpha+1} G)$$
holds for all $\alpha \in \mathbb{Q}$.

Theorem 2.2.4. ([DS03] §3 or [Sai80] §3) For any V^+-solution ω the $\partial_\theta-$action takes the form
$$\nabla_{\theta^2\partial_\theta}\omega = \omega(A_0 + \theta A_\infty),$$
where A_0, A_∞ are two constant $\mu \times \mu-$matrices, A_∞ is semisimple and the spectrum of A_∞ is equal to the spectrum $\mathrm{SP}(G, G_0)$.

So by theorem 2.2.4 a V^+-solution ω is a good basis in the sense of [Sai89].

Later we will consider the situation, where we have solved the Birkhoff problem for a family. I.e. let G be a rational bundle on $U_\infty \times \mathrm{Spec}(\mathbb{C}[t])$, such that each fibre $G(t)$ over $t \in \mathbb{C}_t$ is a meromorphic connection. We denote by $G_0(t)$ a lattice at $\infty = \{\theta = 0\}$ of $G(t)$.

Moreover we will have a basis ω of G_0, such that the connection on $G_0(t)$ is given by
$$\nabla_{\theta^2\partial_\theta}\omega = \omega(A_0(t) + \theta A_\infty), \tag{2.1}$$
where the $\mu \times \mu-$matrices are
$$A_0 = \begin{pmatrix} 0 & 0 & 0 & \cdots & 0 & t \\ 1 & 0 & 0 & \cdots & 0 & 0 \\ 0 & 1 & 0 & \cdots & 0 & 0 \\ 0 & 0 & 1 & \cdots & 0 & 0 \\ \vdots & \vdots & \vdots & \cdots & \vdots & \vdots \\ 0 & 0 & 0 & \cdots & 1 & 0 \end{pmatrix}$$

and $A_\infty = \mathrm{diag}(\nu_1, \ldots, \nu_\mu)$.

Proposition 2.2.5. Let ω be a global basis of $G_0(t)$ such that the connection is given by the matrices $A_0(t)$ and A_∞ as above.
If $t = 0$ and $\nu_i - \nu_{i-1} \leq 1$ holds for $i = 2, \ldots, n$, then ω is a V^+−solution and the spectrum of A_∞ is equal to the spectrum of (G, G_0).
If $t \neq 0$ and additionally $\nu_1 - \nu_n \leq 1$ holds, then the same is true.

Proof. The proof is done in the same way as the proof of [GMS09] proposition 4.8. We will use the numbers ν_i to define a filtration $G(t)$, which fulfils the properties of the V−filtration and hence coincides with the V−filtration by uniqueness.

Let $\omega = (\omega_1, \ldots, \omega_\mu)$ be a basis as in the statement of the proposition. We recall from the discussion above proposition 2.2.2. that

$$G^0(t) := \bigoplus_{i=1}^{\mu} \mathbb{C}[\tau]\omega_i$$

is a $\tau\nabla_{\partial_\tau}$−stable lattice and the operator ∂_τ acts on the basis by

$$\begin{aligned} \partial_\tau \omega_i &= -\omega_{i+1} - \nu_i \tau^{-1} \omega_i, \ 1 \leq i \leq \mu - 1 \\ \partial_\tau \omega_\mu &= -t\omega_1 - \nu_\mu \tau^{-1} \omega_\mu. \end{aligned}$$

We put $\deg(\tau^k \omega_i) := \nu_i - k$ for $i = 1, \ldots, \mu$ and define a filtration $\tilde{V}_\bullet G(t)$ of $G(t)$ by :

$$\tilde{V}_\alpha G(t) := \{\sum_{i=1}^{n} c_i \tau^{k_i} \omega_i \in G(t) \mid \max\{\nu_i - k_i\} \leq \alpha\}.$$

This filtration $\tilde{V}(G(t)$ coincides with the V−filtration by definition, if the operator $(\tau\partial_\tau + \alpha)$ is nilpotent on $\mathrm{gr}_\alpha^{\tilde{V}} G(t)$ for all $\alpha \in \mathbb{Q}$ and the filtration is good with respect to the filtration $V_\bullet \mathbb{C}[\tau]\langle\partial_\tau\rangle$, i.e.

$$V_k \mathbb{C}[\tau]\langle\partial_\tau\rangle \tilde{V}_l G(t) \subset \tilde{V}_{k+l} G(t)$$

and this is an equality for almost all k, l.

Let $\tau^{k_l}\omega_l$ be in $\tilde{V}_\alpha G(t)$. For the τ−action we have by definition $\tau_\tau,^{k_l} \omega_l \in \tilde{V}_{\alpha-1} G(t)$. For the ∂_τ−action we have

$$\partial_\tau \tau^{k_l} \omega_l = k_l \tau^{k_l - 1} \omega_l + \tau^{k_l} \begin{cases} -\omega_{l+1} - \nu_l \tau^{-1} \omega_l & 1 \leq l \leq \mu - 1 \\ -t\omega_1 - \nu_\mu \tau^{-1} \omega_\mu & . \end{cases} \tag{2.2}$$

So we conclude for the degree of $\partial_\tau \tau^{k_l} \omega_l$:

$$\deg(\partial_\tau \tau^{k_l} \omega_l) = \max\{\nu_l - (k_l - 1), \nu_{l+1} - k_l, \nu_l - (k_l - 1)\},$$

where we have set $\nu_{\mu+1} = \nu_1$.
We conclude from the assumption $\nu_{l+1} - \nu_l \leq 1$ for $l \neq \mu$ that we have $\nu_{l+1} - k_l \leq \alpha + 1$ and from the assumption $\nu_1 - \nu_\mu \leq 1$ that we have $\nu_1 - k_\mu \leq \alpha + 1$ for $t \neq 0$. Hence we have $\partial_\tau \tau^{k_l} \omega_l \in \tilde{V}_{\alpha+1} G(t)$. Because τ acts bijective on $G(t)$, the filtration $\tilde{V} G(t)$.
In order to see that the operator $(\tau\partial_\tau + \alpha)$ is nilpotent on $gr_\alpha^{\tilde{V}} G$, we set $\omega_{n+1} := t \cdot \omega_1$ and $\nu_{n+1} := \nu_1$. Because the degree of

$$
\begin{aligned}
(\tau\partial_\tau + \alpha)\tau^{k_i}\omega_i &= (\alpha - (\nu_i - k_i))\tau^{k_i}\omega_i - \tau^{k_i+1}\omega_{i+1} \\
&= -\tau^{k_i+1}\omega_{i+1}
\end{aligned}
$$

is bounded by $\nu_{i+1} - k_i - 1$, the claim follows from the assumption $\nu_{i+1} \leq \nu_i + 1$. So $\tilde{V}_\bullet G(t)$ and the Kashiwara-Malgrange filtration $V_\bullet G(t)$ coincide.
Because we have $G_0(t) \cap G^0(t) = \oplus_{i=1}^{\mu} \mathbb{C}\omega_i$ by construction,

$$
G_0(t) = (G_0(t) \cap G^0(t)) \oplus \tau^{-1} G_0(t)
$$

holds and ω is by definition a $V-$solution. Moreover we have

$$
(\tau\partial_\tau + \alpha)(\bigoplus_{i=1}^{\mu} \mathbb{C}\omega_i \cap V_\alpha G(t)) \subset (\bigoplus_{i=1}^{\mu} \mathbb{C}\omega_i \cap V_{<\alpha} G(t)) \oplus \tau(\bigoplus_{i=1}^{\mu} \mathbb{C}\omega_i \cap V_{\alpha+1} G(t)),
$$

i.e. ω is a V^+-solution, because of

$$
(\tau\partial_\tau + \alpha)(\sum_{i=1}^{\mu} c_i\omega_i) = \sum_{i=1}^{\mu} (\alpha - \nu_i) c_i \omega_i - \sum_{i=1}^{\mu} \tau c_i \omega_{i+1},
$$

where we have set $\omega_{n+1} := t \cdot \omega_1$.

That the $(\nu_i)_{i=1,\ldots,\mu}$ are equal to the spectrum of (G, G_0), follows from the precise description of the $V-$filtration, namely :

$$
\begin{aligned}
V_\alpha G(t) \cap G_0(t) &= \{\sum g_i(\tau^{-1})\omega_i \mid \deg(g_i) \leq \alpha - \nu_i\} \\
V_\alpha G(t) \cap \tau^{-1} G_0(t) &= \{\sum g_i(\tau^{-1})\omega_i \mid 1 \leq \deg(g_i) \leq \alpha - \nu_i\} \\
V_{<\alpha} G(t) \cap G_0(t) &= \{\sum g_i(\tau^{-1})\omega_i \mid \deg(g_i) < \alpha - \nu_i\},
\end{aligned}
$$

where deg denotes the usual degree of the polynomial $g_i \in \mathbb{C}[\tau^{-1}]$. Now the claim follows from

$$
\frac{V_\alpha G(t) \cap G_0(t)}{V_\alpha G(t) \cap \tau^{-1} G_0(t) + V_{<\alpha} G(t) \cap G_0(t)} = \oplus_{\nu_i = \alpha} \mathbb{C}\omega_i.
$$

□

Moreover it was shown in [GMS09] that the assumptions on the entries ν_i of the matrix A_∞ in proposition 2.2.5 are satisfied after applying a base change to a basis ω as in (2.1). The precise statement is the following:

Lemma 2.2.6. ([GMS09] Lemma 4.10-4.11) Let $\omega^{(1)}$ be a solution to the Birkhoff problem, i.e. $\nabla_{\theta^2\partial_\theta}\omega^{(1)} = \omega^{(1)}(A_0 + \theta A_\infty)$ with A_0 and A_∞ as in (2.1), then there exists bases $\omega^{(2)}$ and $\omega^{(3)}$ such that $\omega^{(2)}$ is a V^+−solution for $t = 0$ and $\omega^{(3)}$ is a V^+−solution for $t \neq 0$.

Proof. The statements follow from an application of two algorithms to the basis $\omega^{(1)}$. Because we will need these algorithms later, we briefly recall them. But we refer to the proofs of [GMS09] lemma 4.10 resp. lemma 4.11 that these algorithms terminate.
Algorithm 1 :
Whenever there is $i \in \{2, \dots, n\}$ with $\nu_i - \nu_{i-1} > 1$, we set

$$\widetilde{\omega}_i{}^{(1)} := \omega_i^{(1)} + \theta^{-1}(\nu_i^{(1)} - \nu_{i-1}^{(1)} - 1)\omega_{i-1}^{(1)}$$

and $\widetilde{\omega}_l^{(1)} := \omega_l^{(1)}$ for $l \neq i$.
It follows that in the basis $\widetilde{\omega}$ we have $\theta^2\partial_\theta\widetilde{\omega} = \widetilde{\omega}(A_0 + \theta\tilde{A}_\infty)$ with $\tilde{A}_\infty = \mathrm{diag}(\tilde{\nu}_1, \dots, \tilde{\nu}_n)$, where

$$\begin{aligned} \tilde{\nu}_i &= \nu_{i-1} + 1 \\ \tilde{\nu}_{i-1} &= \nu_i - 1 \end{aligned}$$

and $\tilde{\nu}_j = \nu_j$. Then we restart algorithm 1 with input $\widetilde{\omega}$.
When the inequalities $\nu_i - \nu_{i-1} \leq 1$ hold for all $i = 2, \dots, n$, we put $\omega^{(2)} := \widetilde{\omega}$.
Algorithm 2 :
We first run algorithm 1 with input $\omega^{(1)}$ to obtain the output $\omega^{(2)}$. If we have $\nu_1 - \nu_n > 1$, we set

$$\begin{aligned} \widetilde{\omega}_1^{(2)} &:= t\omega_1^{(2)} + \theta^{-1}(\nu_1^{(2)} - \nu_n^{(2)} - 1)\omega_n^{(2)} \\ \widetilde{\omega}_i^{(2)} &:= t\omega_i^{(2)}, \ \forall i \neq 1. \end{aligned}$$

It follows that in the basis $\widetilde{\omega}^{(2)}$ we have $\theta^2\partial_\theta(\widetilde{\omega}^{(2)}) = \widetilde{\omega}^{(2)}(A_0 + \theta^{-1}\tilde{A}_\infty)$ with $\tilde{A}_\infty = \mathrm{diag}(\tilde{\nu}_1^{(2)}, \dots, \tilde{\nu}_n^{(2)})$, where

$$\begin{aligned} \tilde{\nu}_1^{(2)} &= \nu_n^{(2)} + 1 \\ \tilde{\nu}_n^{(2)} &= \nu_1^{(2)} - 1 \end{aligned}$$

and $\tilde{\nu}_i^{(2)} = \nu_i^{(2)}$ for all $i \neq 1, n$. Then we run algorithm 2 with the input $\widetilde{\omega}^{(2)}$. If $\nu_1 - \nu_n \leq 1$ we put $\omega^{(3)} := \widetilde{\omega}$. □

Chapter 3

Logarithmic vector fields and the logarithmic de Rham complex

In [Sai80] K. Saito associated to a reduced hypersurface D in a complex manifold the sheaves of logarithmic vector fields and logarithmic differential forms along D. In the case that D is the union of smooth subvarieties which are normal crossing, such notions already existed (see for instance [Del70] and [Kat71]).
We will study these modules for a divisor $D \subset V = \mathbb{A}^n_{\mathbb{C}}$ defined by a not necessarily reduced but homogeneous equation h.
We will see later in chapter 4 and 5 that these logarithmic forms appear naturally as coefficients of a Gauss-Manin system of a certain family and that the modules of logarithmic vector fields can be identified with the modules of certain relative de Rham cohomology groups of this family.

We denote by $x_1, \dots, x_n$ coordinates on V and by $\partial_i = \partial_{x_i}$ the corresponding partial derivatives.

3.1 Logarithmic forms and vector fields

In this section we will give the definitions of logarithmic vector fields and differential forms along a homogeneous divisor $D \subset V$ with equation h. We will compare the reduced with the non reduced case and therefore we denote by h^{red} the reduced equation and by D^{red} its reduced divisor.

Definition 3.1.1. A vector field $\xi \in \mathrm{Der}_V$ is called *logarithmic* (along D), if $\xi(h) \in \mathcal{O}_V h$.

We denote by $\mathrm{Der}_V(-log\ D)$ the $\mathcal{O}_V$−module of logarithmic vector fields along D and will often drop the index V.

One can show that $\mathrm{Der}(-log\ D)$ is a $\mathcal{O}_V$−coherent submodule of Der_V, which is closed under the Lie bracket $[\cdot,\cdot]$.

To compare the modules of logarithmic vector fields of D and D^{red} we use a different notation for a moment. For a regular function $g \in \mathcal{O}_V$ we denote by D_g its divisor with possible non-reduced structure.

Lemma 3.1.2. ([GS06] lemma. 3.4) For $f, g \in \mathcal{O}_V$ we have

$$\mathrm{Der}(-log\ D_{gf}) = \mathrm{Der}(-log\ D_g) \cap \mathrm{Der}(-log\ D_f).$$

Proof. Let $g \in \mathcal{O}_V$ a regular function and $\prod_{i=1}^k g_i^{e_i}$ be the decomposition into irreducible factors and $\xi \in \mathrm{Der}_V$. We conclude from

$$\xi(g) = \sum_{i=1}^{k} e_i \xi(g_i) \cdot \frac{g}{g_i}$$

that $\mathrm{Der}(-log\ D_g) = \cap_{i=1}^k \mathrm{Der}(-log\ D_{g_i})$. □

Corollary 3.1.3. For a divisor $D \subset V$ we have

$$\mathrm{Der}(-log\ D) = \mathrm{Der}(-log\ D^{red}).$$

Therefore we will not distinguish in the following between $\mathrm{Der}(-log\ D)$ and $\mathrm{Der}(-log\ D^{red})$.

So far we have not used the homogeneity property of D. It yields a specific decomposition of $\mathrm{Der}(-log\ D)$ due to the existence of an Euler field.

Definition 3.1.4. We call an *Euler field of weight* $r \in \mathbb{C}^*$ a vector field $E \in \mathrm{Der}_V(-log\ D)$ such that $E(h) = rh$.

Remark 3.1.5. In general an Euler field does not need to exists, but as h is homogeneous, $E := \sum_{i=1}^n x_i \partial_i$ is an Euler field of weight $r = \deg(h)$. The submodule

$$\mathrm{Der}(-log\ h) := \{\xi \in \mathrm{Der}(-log\ D) \mid \xi(h) = 0\}$$

describes all vector fields, which are tangent to all fibres of h and due to the presence of an Euler field the formula

$$\xi = \frac{\xi(h)}{E(h)} E + (\xi - \frac{\xi(h)}{E(h)} E)$$

defines a direct sum decomposition

$$\mathrm{Der}(-log\ D) = \mathcal{O}_V E \oplus \mathrm{Der}(-log\ h).$$

Indeed let $\xi \in \mathrm{Der}(-log\ D)\backslash\{0\}$ and

$$\eta \in (\mathcal{O}_V \frac{\xi(h)}{E(h)} E) \cap (\mathcal{O}_V(\xi - \frac{\xi(h)}{E(h)} E))$$

Then there exists $c_1, c_2 \in \mathcal{O}_V$, such that

$$\eta = c_1 \cdot \frac{\xi(h)}{E(h)} E = c_2 \cdot (\xi - \frac{\xi(h)}{E(h)} E).$$

In the case $\xi(h) = 0$ we have $0 = c_2\xi$, so $c_2 = 0$ and hence $\eta = 0$.
In the case $\xi(h) \neq 0$, e.g. $\xi(h) = gh$ for some $g \in \mathcal{O}_V\backslash\{0\}$, we have

$$\eta = c_1 \cdot \frac{gh}{rh} E = c_2 \cdot (\xi - \frac{gh}{rh} E),$$

for $r = \deg(h)$. If we apply η to h, we get

$$\begin{aligned} c_1 \frac{g}{r} E(h) &= c_2(\xi - \frac{g}{r} E)(h) \\ c_1 gh &= c_2(gh - gh) = 0. \end{aligned}$$

So we conclude $c_1 = 0$ and hence $\eta = 0$.

Corollary 3.1.6. Let$D$ be a divisor with homogeneous defining equation $h \in \mathcal{O}_V$. Let E be an Euler field of weight r, then there is a decomposition

$$\mathrm{Der}(-log\ D) = \mathcal{O}_V \oplus \mathrm{Der}(-log\ h).$$

Definition 3.1.7. We call a rational $k-$form ω on V with poles along D *logarithmic (along D)*, if $h\omega$ and $dh \wedge \omega$ are regular forms on V. We denote by $\Omega^k_V(log\ D)$ the $\mathcal{O}_V-$module of $k-$logarithmic forms on V.

Lemma 3.1.8. A $k-$form ω is logarithmic if and only if $h\omega \in \Omega^k_V$ and $hd\omega \in \Omega^{k+1}_V$. In particular we have

$$\Omega^k_V(log\ D^{red}) \subset \Omega^k_V(log\ D).$$

Proof. Let ω be a rational $k-$form such that $h\omega \in \Omega^k_V$. We conclude from

$$d(h\omega) = dh \wedge \omega - hd\omega,$$

that $dh \wedge \omega$ is regular if and only if $hd\omega$ is regular. So in particular for any $\omega \in \Omega^k_V(log\ D^{red})$ the forms $h\omega$ and $hd\omega$ are regular. □

As in [Sai80] §1 one concludes that $\Omega^k_V(log\ D)$ are coherent $\mathcal{O}_V-$modules and $\Omega^n_V(log\ D) = \frac{1}{h}\Omega^n$. In particular $\Omega^n_V(log\ D)$ is a free $\mathcal{O}_V-$modul of rank 1 generated by $\frac{dx_1\wedge\ldots\wedge dx_n}{h}$ and $\Omega^n_V(log\ D^{red})$ is freely generated by $\frac{dx_1\wedge\ldots\wedge dx_n}{h^{red}}$.

Remark 3.1.9. In the case of an reduced equation we recall from [Sai80] that $\oplus_{k=0}^n\Omega^k_V(log\ D^{red})$ is an $\mathcal{O}_V-$exterior algebra and that $\Omega^1_V(log\ D^{red})$ and $\mathrm{Der}(-log\ D^{red})$ are reflexiv and mutually dual to each other. Therefore Saito showed that the exterior derivative d induces a differential

$$d\colon \Omega^k_V(log\ D^{red}) \to \Omega^{k+1}_V(log\ D^{red})$$

and for $\xi \in \mathrm{Der}(-log\ D)$ the interior product extends to

$$i_\xi\colon \Omega^k_V(log\ D^{red}) \to \Omega^{k-1}_V(log\ D^{red}).$$

This way one obtains a well defined inner product

$$\begin{array}{rcl} \mathrm{Der}(-log\ D^{red}) \times \Omega^1_V(log\ D^{red}) & \to & \mathcal{O}_V \\ (\xi,\omega) & \mapsto & \omega(\xi). \end{array}$$

The statement for the exterior derivate follows directly from the definitions, but for the interior product Saito used an equivalent description of a logarithmic $k-$form ω in the reduced case, namely :
There exists a non-zerodivisor $g \in \mathcal{O}_V$, $\sigma \in \Omega^{k-1}_V$ and $\eta \in \Omega^k_V$, such that

1. $\dim_{\mathbb{C}}(D^{red} \cap \mathcal{V}(g)) \leq n-2$
2. $g\omega = \frac{dh^{red}}{h^{red}} \wedge \sigma + \eta$.

So if $\xi \in \mathrm{Der}(-log\ D)$ and $\omega \in \Omega^k(log\ D^{red})$ satisfy $g\omega = \frac{dh}{h} \wedge \sigma + \eta$, for some non-zerodivisor $g \in \mathcal{O}_V$ on D and some regular forms σ, η, the claim follows from

$$gi_\xi(\omega) = i_\xi(g\omega) = i_\xi(\frac{dh}{h}) \wedge \sigma - \frac{dh}{h} \wedge i_\xi(\sigma) + i_\xi(\eta).$$

Indeed, $\xi(\frac{dh}{h}), i_\xi(\sigma)$ and $i_\xi(\eta)$ are regular and only $\frac{dh}{h} \wedge i_\xi(\sigma)$ may not lie in Ω^{k-1}_V, but it lies by definition in $\Omega^{k-1}(log\ D)$.

Remark 3.1.10. In case of a non-reduced equation the exterior product extends from regular to logarithmic $k-$forms by definition and also the interior product extends, as we will see later. But this can not be done with Saito's equivalent description of a logarithmic $k-$form, because this description, i.e. [Sai80] lemma 1.1, is not necessarily true for a non

reduced equation h.
We will see that in general $\oplus_{i=0}^{n}\Omega_V^i(log\ D)$ is not an exterior $\mathcal{O}_V-$algebra and moreover Der$(-log\ D)$ and $\Omega_V^1(log\ D)$ are not reflexiv and dual to each other. Before we are going to prove these statements, we look at two examples first.

Example 3.1.11. For $V=\mathbb{C}^1$ and $h(x)=x^2$ we have

$$\mathcal{O}_V\frac{dx}{x}=\Omega^1(log\ D^{red})\subsetneq\mathcal{O}_V\frac{dx}{x^2}=\Omega^1(log\ D).$$

Example 3.1.12. For $V=\mathbb{C}^2$ and $h=x_1x_2^2$ the $1-$forms $\omega_1=\frac{dx_1}{x_1x_2}$ and $\omega_2=\frac{dx_2}{x_2}$ are in $\Omega_V^1(log\ D)$, but $\omega_1\wedge\omega_2=\frac{dx_1\wedge dx_2}{x_1x_2^3}$ is not in $\Omega_V^2(log\ D)$.

Remark 3.1.13. We conclude from example 3.1.11 that the inclusion of lemma 3.1.8 is not an equality in general. From example 3.1.12 we conclude that in general $\Omega_V^k(log\ D)\neq\wedge^k\Omega_V^1(log\ D)$.
In particular $\bigoplus_{i=0}^{n}\Omega_V^i(log\ D)$ is not an $\mathcal{O}_V-$exterior algebra.

Remark 3.1.14. In general even in the reduced case $\Omega_V^1(log\ D)$ and Der$(-log\ D)$ are not locally free $\mathcal{O}_V-$sheaves, although both sheaves coincide outside D with the free sheaves Ω_V^1 and Der$_V$. But in the reduced case we have the following criterion for freeness:

Proposition 3.1.15. ([Sai80] Theorem 1.8) Let $D\subset V$ be a reduced divisor, then we have:

1. $\Omega_V^1(log\ D)$ is $\mathcal{O}_V-$free if and only if $\wedge^n\Omega_V^1(log\ D)=\Omega_V^n(log\ D)$.

2. Der$(-log\ D)$ is $\mathcal{O}_V-$free if and only if there exists elements $\xi_1,\dots,\xi_n$ in Der$(-log\ D)$ with $\xi_i=\sum_j g_{ij}(x)\partial_{x_j}$, such that the determinant $\det(g_{ij}(x))$ is a unit multiple of h.

The last criterion is known as *Saito's criterion* and a divisor D fulfilling it is a free divisor. If moreover the $g_{ij}(x)$ are linear, then the divisor D is a linear free divisor. Furthermore if the divisor is free, then by [Sai80] (1.8)

$$\Omega_V^k(log\ D)=\wedge^k\Omega_V^1(log\ D).$$

We denote by M^* the $\mathcal{O}_V-$dual of a $\mathcal{O}_V-$module M. Then corollary 3.1.3 and Saito's result, that Der$(-log\ D^{red})$ and $\Omega_V^1(log\ D^{red})$ are reflexiv and dual to each other, directly imply:

Lemma 3.1.16. We have $(\mathrm{Der}(-log\ D))^* = \Omega^1(log\ D^{red})$.
But this does not hold in general, as by example 3.1.11 we have

$$(\mathrm{Der}(-log\ D)^* = \Omega^1_V(log\ D^{red}) \neq \Omega^1(log\ D).$$

Let us deal with the statement about the interior product next. We denote by $\Omega^k_V(*D)$ the module of differential $k-$forms which are rational along D, i.e. $\Omega^k_V(*D) = \mathcal{O}_V(*D) \otimes \Omega^k_V$.

Lemma 3.1.17. For $\xi \in \mathrm{Der}(-log\ D)$ the interior product

$$\iota_\xi \colon \Omega^k_V(*D) \to \Omega^{k-1}_V(*D)$$

restricts to

$$\iota_\xi \colon \Omega^k_V(log\ D) \to \Omega^{k-1}_V(log\ D)$$

if and only if $\xi \in \mathrm{Der}(-log\ D)$.

Proof. By definition a rational $k-$form $\omega \in \Omega^k_V(*D)$ is logarithmic if and only if (1): $h\omega \in \Omega^k_V$ and (2): $dh \wedge \omega \in \Omega^{k+1}_V$ hold.
In particular we have by (1) that the module $\Omega^k_V(log\ D)$ is a submodule of $\frac{1}{h}\Omega^k_V$. Hence by restriction we obtain the map

$$\iota_\xi \colon \Omega^k_V(log\ D) \to \Omega^{k-1}_V(*D)$$

Hence to prove the claim of this lemma, we need to understand, when the image of ι_ξ lies in $\Omega^{k-1}_V(log\ D)$.
For all $\omega \in \Omega^k_V(log\ D)$ the image $\iota_\xi(\omega)$ satisfies condition (1), because ι_ξ is $\mathcal{O}_V-$linear. For condition (2) we consider the identity

$$dh \wedge \iota_\xi(\omega) = -\iota_\xi(dh \wedge \omega) + \xi(h)\omega$$

and note that $\iota_\xi(dh \wedge \omega)$ is always a regular differential form, because ω is logarithmic. But $\xi(h)\omega$ is regular for all $\omega \in \Omega^k_V(log\ D)$ if and only if $\xi \in \mathrm{Der}(-log\ D)$. □

By the proof of lemma 3.1.17 we have an $\mathcal{O}_V-$linear map

$$\begin{array}{rcl}\phi_\omega \colon \mathrm{Der}(-log\ D) & \to & \Omega^{k-1}_V(log\ D) \\ \xi & \mapsto & \iota_\xi(\omega)\end{array}$$

for $\omega \in \Omega^k_V(log\ D)$. In the special case $k = n$ and $\omega = \frac{vol}{h}$ we get

$$\begin{array}{rcl}\phi_{\frac{vol}{h}} \colon \mathrm{Der}(-log\ D) & \to & \Omega^{n-1}_V(log\ D) \\ \xi & \mapsto & \iota_\xi(\dfrac{vol}{h}).\end{array}$$

Using local coordinates we write $\xi = \sum \xi_i \partial_i$ and get

$$\phi_{\frac{vol}{h}}(\xi) = \iota_\xi(\frac{vol}{h}) = \sum \xi_i \iota_{\partial_i}(\frac{vol}{h}) = \sum \frac{(-1)^{i-1}\xi_i}{h}\hat{dx_i},$$

where $\hat{dx_i} = dx_1 \wedge \ldots \wedge dx_{i-1} \wedge dx_{i+1} \wedge \ldots \wedge dx_n$. As $\xi \in \mathrm{Der}(-log\ D)$ we have $\xi(h) = \sum \xi_i \partial_i h \in (h)$ and hence $dh \wedge \iota_\xi(\frac{vol}{h}) = \sum \frac{\xi \partial_i h}{h} vol$ is a regular $n-$form. On the other hand we can write any $\omega \in \Omega_V^{n-1}(log\ D)$ as $\omega = \sum \frac{\omega_i}{h}$.
Because $dh \wedge \omega = \sum \frac{(-1)^{i-1}\omega_i \partial_i h}{h} vol$ is a regular $n-$form, we conclude that $\sum (-1)^{i-1}\omega_i \partial_i \in \mathrm{Der}(-log\ D)$. This way we obtain an $\mathcal{O}_V-$linear map

$$\begin{aligned} \psi\colon \Omega_V^{n-1}(log\ D) &\rightarrow \mathrm{Der}(-log\ D) \\ \sum \frac{\omega_i}{h} &\mapsto \sum (-1)^{i-1}\omega_i \partial_i, \end{aligned}$$

which is inverse to $\phi_{\frac{vol}{h}}$. If we replace h by h^{red} in the above discussion, we obtain :

Proposition 3.1.18. The multiplication with $\frac{h}{h^{red}}$ defines an $\mathcal{O}_V$-isomorphism from $\Omega_V^{n-1}(log\ D)$ to $\Omega_V^{n-1}(log\ D^{red})$ with inverse given by the multiplication with $\frac{h^{red}}{h}$.
Moreover both modules are isomorphic to $\mathrm{Der}(-log\ D)$.

The statement of the proposition is not a surprise, as $\mathrm{Der}(-log\ D)$ can be defined by the first syzygy-module[1] of $(\partial_1 h, \ldots, \partial_n h, h)$ and $\Omega_V^{n-1}(log\ D)$ by the first syzygy-module[2] of $(\partial_1 h, -\partial_2 h, \ldots, (-1)^{n-1}\partial_n h, h)$.

Corollary 3.1.19. Let $D \subset V$ be a divisor defined by an homogeneous polynomial h. Then we have natural isomorphisms

$$\Omega_V^i(log\ D^{red}) \cong \Omega_V^i(log\ D), \text{ for } i = n-1, n.$$

3.2 The relative de Rham complex

In the sequel we are interested in differential forms relative to the fibres of h, which are logarithmic along D.

[1]Precisely the coefficients of $\sum_i \xi_i \partial_i \in \mathrm{Der}(-log\ D)$ are given by the projection to the first $n-$components of the syzygy module.

[2]Precisely the coefficients of $\sum_i \frac{\omega_i}{h}\hat{dx_i} \in \Omega_V^{n-1}(log\ D)$ are given by the projection to the first $n-$components of the syzygy module.

Definition 3.2.1. The *relative logarithmic de Rham complex* is the complex $(\Omega^\bullet_{rel}(log\ D), d)$, where the terms are given by

$$\Omega^k_{rel}(log\ D) := \frac{\Omega^k(log\ D)}{\frac{dh}{h} \wedge \Omega^{k-1}(log\ D)}$$

and the differential is defined by $d[\omega] := [d\omega]$.
In the same way we define the relative (rational) de Rham complex $(\Omega^\bullet_{rel}(*D), d)$.

Let E be an Euler field of weight $r \in \mathbb{C}^*$, i.e. $\iota_E(dh) = rh$, and

$$\begin{aligned} \phi_{\frac{vol}{h}} : \mathrm{Der}(-log\ D) &\to \Omega^{n-1}_V(log\ D) \\ \xi &\mapsto \iota_\xi(\frac{vol}{h}) \end{aligned}$$

be the isomorphism $\xi \mapsto \iota_\xi(\frac{vol}{h})$ from the previous section. By asking, how the decomposition $\mathrm{Der}(-log\ D) = \mathcal{O}_V E \oplus \mathrm{Der}(-log\ h)$ carries over to $\Omega^{n-1}_V(log\ D)$ under $\phi_{\frac{vol}{h}}$, we get a description of the top relative logarithmic de Rham module.

Lemma 3.2.2. The module $\Omega^{n-1}_{rel}(log\ D)$ is freely generated by

$$[\alpha] = [\iota_E(\frac{vol}{h})].$$

Proof. First we deduce two identities in $\Omega^{n-1}_{rel}(log\ D) = \frac{\Omega^{n-1}_V(log\ D)}{\frac{dh}{h} \wedge \Omega^{n-2}_V(log\ D)}$.
Form $\iota_E(dh \wedge \omega) = rh\omega - dh \wedge \iota_E(\omega)$ we conclude

$$[\omega] = [\frac{1}{r}\iota_E(\frac{dh}{h} \wedge \omega)]$$

and from $0 = \iota_\xi(dh \wedge vol) = \xi(h)vol - dh \wedge \iota_\xi(vol)$ we conclude

$$dh \wedge \iota_\xi(vol) = \xi(h)vol.$$

Next we consider the composition

$$\pi \circ \phi_{\frac{vol}{h}} : \mathrm{Der}(-log\ D) \to \Omega^{n-1}_{rel}(log\ D)$$

of the canonical projection $\pi \colon \Omega^{n-1}_V(log\ D) \to \Omega^{n-1}_{rel}(log\ D)$ and the isomorphism $\phi_{\frac{vol}{h}} : \mathrm{Der}(-log\ D) \to \Omega^{n-1}_V(log\ D)$.

We calculate in $\Omega^{n-1}_{rel}(log\ D)$

$$\begin{aligned}
\pi \circ \phi_{\frac{vol}{h}}(\xi) &= [\iota_\xi(\frac{vol}{h})] \\
&= [\frac{1}{r}\iota_E(\frac{dh}{h} \wedge \iota_\xi(\frac{vol}{h})] \\
&= [\frac{1}{r}\iota_E(\frac{\xi(h)}{h}\frac{vol}{h})] \\
&= [\frac{1}{r}\frac{\xi(h)}{h}\alpha].
\end{aligned}$$

and conclude $\mathrm{Der}(-log\ h) \subset \ker(\pi \circ \phi_{\frac{vol}{h}})$.
Obviously we have $\pi \circ \phi_{\frac{vol}{h}}(E) = [\alpha]$.
Suppose we know that $[\alpha] \neq 0$ in $\Omega^{n-1}_{rel}(log\ D)$, then we conclude from the decomposition of $\mathrm{Der}(-log\ D) = \mathcal{O}_V E \oplus \mathrm{Der}(-log\ h)$ that $\ker(\pi \circ \phi_{\frac{vol}{h}}) = \mathrm{Der}(-log\ h)$ and $\mathrm{im}(\pi \circ \phi_{\frac{vol}{h}}) = \mathcal{O}_V[\alpha] \cong \mathcal{O}_V E$.
Suppose $\alpha = \iota_E(\frac{vol}{h}) = \frac{dh}{h} \wedge \eta$ for some $\eta \in \Omega^{n-2}_V(log\ D)$. Because of

$$0 = \iota_E \circ \iota_E(\frac{vol}{h}) = \iota_E(\frac{dh}{h} \wedge \eta) = r\eta - \frac{dh}{h} \wedge \iota_E(\eta),$$

we have $\eta = \frac{1}{r}\frac{dh}{h} \wedge \iota_E(\eta)$. Hence we get the contradiction

$$\alpha = \iota_E(\frac{vol}{h}) = 0 \in \Omega^{n-1}_V(log\ D).$$

Indeed we have $\alpha = \iota_E(\frac{vol}{h}) \neq 0$ in general. To see this we write $E = \sum E_i \partial_i$. Then we have

$$0 = \iota_E(\frac{vol}{h}) = \frac{\sum(-1)^{i-1}E_i\widehat{dx_i}}{h},$$

where $\widehat{dx_i} = dx_1 \wedge \ldots \wedge dx_{i-1} \wedge dx_{i+1} \wedge \ldots \wedge dx_n$. Because $\frac{\widehat{dx_i}}{h}$ is an $\mathcal{O}_V-$basis of $\frac{1}{h}\Omega^{n-1}_V$, we conclude $E_i = 0$ for all i, which contradicts the assumption $E(h) = rh$. □

Corollary 3.2.3. We have an exact sequence of $\mathcal{O}_V-$modules

$$0 \to \mathrm{Der}(-log\ h) \xrightarrow{\phi} \Omega^{n-1}_V(log\ D) \xrightarrow{\pi} \Omega^{n-1}_{rel}(log\ D) \cong \mathcal{O}_V[\alpha] \to 0,$$

which splits, i.e. $\Omega^{n-1}_V(log\ D) \cong \mathrm{Der}(-log\ h) \oplus \Omega^{n-1}_{rel}(log\ D)$.
We denote by $\ker^k(\frac{dh}{h}\wedge)$ the kernel of the $\mathcal{O}_V-$homomorphism

$$\begin{aligned}
\frac{dh}{h}: \Omega^k_V(log\ D) &\to \Omega^{k+1}_V(log\ D) \\
\omega &\mapsto \frac{dh}{h} \wedge \omega.
\end{aligned}$$

Lemma 3.2.4. The short exact sequence

$$0 \to \ker^k(\frac{dh}{h}\wedge) \to \Omega^k_V(log\ D) \to \frac{dh}{h} \wedge \Omega^k_V(log\ D) \to 0$$

splits for all k, i.e. $\Omega^k_V(log\ D) \cong \ker^k(\frac{dh}{h}\wedge) \oplus \frac{dh}{h} \wedge \Omega^k_V(log\ D)$.
In the special case $k = n-2$ we have

$$\Omega^{n-2}_V(log\ D) \cong \mathrm{Der}(-log\ h) \oplus \ker^{n-2}(\frac{dh}{h}\wedge).$$

Proof. The $\mathcal{O}_V$−homomorphism

$$\begin{aligned} \psi\colon \frac{dh}{h} \wedge \Omega^k_V(log\ D) &\to \Omega^k_V(log\ D) \\ \frac{dh}{h} \wedge \omega &\mapsto \frac{1}{r}\iota_E(\frac{dh}{h} \wedge \omega), \end{aligned}$$

satisfies $\frac{dh}{h} \wedge \psi(\frac{dh}{h} \wedge \omega) = \frac{dh}{h} \wedge \omega$, so the short exact sequence splits.
In the special case $k = n-2$ we have by corollary 3.2.3

$$\mathrm{Der}(-log\ h) \cong \ker(\pi) = \frac{dh}{h} \wedge \Omega^{n-2}_V(log\ D)$$

and hence $\Omega^{n-2}_V(log\ D) \cong \mathrm{Der}(-log\ h) \oplus \ker^{n-2}(\frac{dh}{h}\wedge)$. □

We note that the splitting in the special case from above is given by

$$\begin{aligned} \mathrm{Der}(-log\ h) &\to \Omega^{n-2}_V(log\ D) \\ \xi &\mapsto \frac{1}{r}\iota_E\iota_\xi(\frac{vol}{h}). \end{aligned}$$

We will finish this section by a result, which will be important, when we study the Gauss-Manin system of a pair (h, f). But first we need the following lemma:

Lemma 3.2.5. For all $\xi \in \mathrm{Der}(-log\ h)$ and all $g \in \mathcal{O}_V$ we have in $\Omega^{n-1}_{rel}(log\ D)$:

$$\iota_\xi(dg \wedge \alpha) = 0.$$

In particular we have $\xi(g)\alpha = dg \wedge \iota_\xi(\alpha)$ in $\Omega^{n-1}_{rel}(log\ D)$.

Proof. First we deduce form $0 = \iota_E(dg \wedge \frac{vol}{h}) = E(g)\frac{vol}{h} - dg \wedge \iota_E(\frac{vol}{h})$ that we have

$$E(g)\frac{vol}{h} = dg \wedge \iota_E(\frac{vol}{h})$$

and in the special case of $g = h$ we have

$$\frac{vol}{h} = \frac{dh}{rh} \wedge \alpha.$$

Hence for $\xi \in \mathrm{Der}(-log\ h)$ we have in $\Omega_V^{n-1}(log\ D)$

$$\begin{aligned}
\iota_\xi(dg \wedge \alpha) &= \iota_\xi(dg \wedge \iota_E(\frac{vol}{h})) \\
&= \iota_\xi(E(g)\frac{vol}{h}) \\
&= \iota_\xi(E(g)\frac{dh}{rh} \wedge \alpha) \\
&= \frac{E(g)}{r}\left(\iota_\xi(\frac{dh}{h}) \wedge \alpha - \frac{dh}{h} \wedge \iota_\xi(\alpha)\right) \\
&= \frac{-E(g)}{r}\frac{dh}{h} \wedge \iota_\xi(\alpha).
\end{aligned}$$

Hence $\iota_\xi(dg \wedge \alpha)$ represents the class $[0]$ in $\Omega_{rel}^{n-1}(log\ D) = \frac{\Omega_V^{n-1}(log\ D)}{\frac{dh}{h} \wedge \Omega_V^{n-2}(log\ D)}$. ☐

Proposition 3.2.6. Let $f \in \mathcal{O}_V$ be a linear form, then

$$\frac{\Omega_{rel}^{n-1}(log\ D)}{df \wedge \Omega_{rel}^{n-2}(log\ D)} \cong \frac{\mathcal{O}_V}{df(\mathrm{Der}(-log\ h))}[\alpha].$$

Proof. By Lemma 3.2.2 we have $\Omega_{rel}^{n-1}(log\ D) \cong \mathcal{O}_V[\alpha]$, therefore we have to identify $df \wedge \Omega_{rel}^{n-2}(log\ D)$ with $df(\mathrm{Der}(-log\ h))$.
Let $\omega \in \Omega_V^{n-2}(log\ D)$. Then there exists by lemma 3.2.4 a $\xi \in \mathrm{Der}(-log\ h)$ and a $\kappa \in \ker^{n-2}(\frac{dh}{h}\wedge)$, such that :

$$\omega = \frac{1}{r}\iota_E\iota_\xi(\frac{vol}{h}) + \kappa.$$

Because of $0 = \iota_E(\frac{dh}{h} \wedge \kappa)$ we have $\kappa = \frac{1}{r}\frac{dh}{h} \wedge \iota_E(\kappa)$ and hence we have in $\Omega^{n-1}_{rel}(log\ D)$:

$$
\begin{aligned}
[df \wedge \omega] &= [df \wedge \frac{1}{r}\iota_E \iota_\xi(\frac{vol}{h}) + df \wedge \kappa] \\
&= [df \wedge \frac{1}{r}\iota_E \iota_\xi(\frac{vol}{h}) + df \wedge \frac{1}{r}\frac{dh}{h} \wedge \iota_E(\kappa)] \\
&= [\frac{1}{r} df \wedge \iota_E \iota_\xi(\frac{vol}{h})] - \frac{1}{r}[\frac{dh}{h} \wedge df \wedge \iota_E(\kappa)] \\
&= -\frac{1}{r}[df \wedge \iota_\xi \iota_E(\frac{vol}{h})] \\
&= -\frac{1}{r}[\iota_\xi(df) \wedge \alpha] \\
&= -\frac{1}{r}[\xi(f)\alpha],
\end{aligned}
$$

where in the 5th step we used lemma 3.2.5. We conclude from the calculation that we have $df \wedge \Omega^{n-2}_{rel}(log\ D) \subset df(\mathrm{Der}(-log\ h))[\alpha]$ under the isomorphism $\Omega^{n-1}_{rel} \cong \mathcal{O}_V[\alpha]$.
On the other hand if $\omega = [g\alpha] \in \Omega^{n-1}_{rel}(log\ D) = \mathcal{O}_V[\alpha]$ lies in the kernel of the canonical projection $\pi \colon \mathcal{O}_V[\alpha] \to \frac{\mathcal{O}_V}{df(\mathrm{Der}(-log\ h))}[\alpha]$, we have by lemma 3.2.5 :

$$
\begin{aligned}
\pi([g\alpha]) &= [df(\xi)\alpha] \\
&= [\iota_\xi(df) \wedge \alpha] \\
&= [df \wedge \iota_\xi(\alpha)].
\end{aligned}
$$

Hence we have $df(\mathrm{Der}(-log\ h)) \subset df \wedge \Omega^{n-2}_{rel}(log\ D)$. □

Chapter 4

Functions on prehomogeneous discriminants and their Milnor fibrations

In this chapter we study hyperplane sections with the fibres of a certain homogeneous polynomial h. In the case that the zero fibre $D = h^{-1}(0)$ is a linear free divisor, this was studied in [GMS09] §3. So let us describe our situation and compare it to the one of [GMS09]:

Let (G, ρ, V) be a prehomogeneous vector space with $\dim(G) = \dim(V)$. We assume that the group G is connected and the singular set is a divisor D. Let h be an defining equation for D. In particular h is a semi-invariant and homogeneous by lemma 1.1.8.
We denote by $n_h = \deg(h)$ its degree and by $\chi_h \in \mathrm{Hom}(G, \mathbb{C}^*)$ its character, i.e. we have $g.h(x) = h(g^{-1}.x) = \chi_h(g)h(x)$ for all $g \in G$. Moreover we denote by $\mathfrak{g}$ the Lie algebra of G and by

$$\mathfrak{a} := \ker(d\chi_h) \subset \mathfrak{g}$$

the kernel of the differential of χ_h. We recall from lemma 1.4.1 that

$$\begin{array}{rcl} \phi\colon \mathfrak{g} & \to & \mathrm{Der}(-log\ D)_0 \\ A & \mapsto & x^T A^T \partial \end{array}$$

is an isomorphism of Lie algebras, where $\mathrm{Der}(-log\ D)_0$ denotes the module of logarithmic vector fields of degree 0.

Definition 4.0.1. We denote by $\tilde{\mathfrak{g}}$ the $\mathcal{O}_V$−submodule of $\mathrm{Der}(-log\ D)$ generated by $\phi(\mathfrak{g})$ and by $\tilde{\mathfrak{a}}$ the $\mathcal{O}_V$−submodule of $\mathrm{Der}(-log\ D)$ generated by $\phi(\mathfrak{a})$.

Obviously $\tilde{\mathfrak{a}}$ is a Lie-subalgebra of Der($-log\ h$) and $\tilde{\mathfrak{g}}$ is a Lie-subalgebra of Der($-log\ D$). By lemma 1.4.1 the inclusions are equalities, i.e. $\tilde{\mathfrak{g}} =$ Der($-log\ D$) and $\tilde{\mathfrak{a}} =$ Der($-log\ h$)), if and only if D is a linear free divisor. We recall that a linear free divisor is a special case of a discriminant D of a prehomogeneous vector space (G, ρ, V), such that

1. G is connected linear algebraic group and $\dim(G) = \dim(V)$.
2. $\rho = \mathrm{id}\colon \mathrm{G} \to \mathrm{GL}(V)$ and
3. $\tilde{h}(x) := \det(d\rho(A_1)x| \dots |d\rho(A_n)x)$ is a reduced equation for D, where $A_1, \dots, A_n$ is a basis of $\mathfrak{g}$.

We will not assume the third condition and consider moreover an arbitrary equation h for D in the following. But from section 4.2 on we will assume that (G, ρ, V) is a regular reductive prehomogeneous vector space.
As the reductivity of G was also assumed in [GMS09] §3 later, this chapter can be seen as a generalization of their results.

In the case that the discriminant D is a linear free divisor, the modules Der($-log\ D$) and Der($-log\ h$) are free and many arguments in [GMS09] depend on these modules and their freeness. But both modules are in general neither free in case of an arbitrary prehomogeneous discriminant D nor if one considers non reduced equations h for D. In order to solve this problem we work with the free submodules $\tilde{\mathfrak{g}}$ and $\tilde{\mathfrak{a}}$ instead.

Furthermore in this chapter we will consider the restriction of an homogeneous $f \in V^*$ to thr fibres $D = h^{-1}(0)$ and $D_t = h^{-1}(t)$ for $t \neq 0$ and assume in this chapter, that every occurring object is defined over $\mathbb{C}$. We recall that the differential $d\alpha_f$ of the orbit map

$$\begin{aligned} \alpha_f\colon G &\to V^* \\ g &\mapsto g.f, \end{aligned}$$

factors by corollary 1.4.3 through $\phi\colon \mathfrak{g} \to$ Der($-log\ D$), $A \mapsto x^T A^T \partial$ and $-df\colon \tilde{g} \to \mathcal{O}_V$, $\xi \mapsto -\iota_\xi(f)$ and that the images $d\alpha_f(\mathfrak{g})$ and $df(\tilde{\mathfrak{g}})$ generate the same $\mathcal{O}_V$-module.

We remark that in [GMS11] and [GS10] discriminants of regular reductive prehomogeneous vector spaces given with equation $\tilde{h}$ where studied too.

4.1 $\mathcal{R}$–finiteness of linear functions on discriminants

Let h be an equation of the discriminant D of a prehomogeneous vector space (G, ρ, V) and $f\colon V \to \mathbb{C}$ be a linear function, which we also consider as a function on the fibres of h.

In [GMS09] several notions of finiteness are considered for the linear function f. They have a deformation theoretic interpretation, that is, they can be considered as finiteness conditions for tangent spaces to orbits of various group actions. The two most important tangent spaces are $df(\mathrm{Der}(-log\ D))$ and $df(\mathrm{Der}(-log\ h))$. We will introduce similar notions of finiteness in definition 4.1.3 below by replacing $df(\mathrm{Der}(-log\ D))$ by $\tilde{\mathfrak{g}}$ and $df(\mathrm{Der}(-log\ h))$ by $\tilde{\mathfrak{a}}$. These notions coincide if D is a linear free divisor.

Lemma 4.1.1. $\tilde{\mathfrak{g}}$ is a free $\mathcal{O}_V$–submodule of $\mathrm{Der}(-log\ D)$.

Proof. Let $A_1, \ldots, A_n$ be a basis of $\mathfrak{g}$ and $s_1, \ldots, s_n \in \mathcal{O}_V$, such that we have $\sum_i s_i x^T A_i^T \partial = 0$. Obviously this holds if and only if the regular functions $S_{ij}(x) := (s_i A_i x)_j$, $j = 1, \ldots, n$, vanish on V. In particular the S_{ij} vanish on the open and dense G–orbit U in V in this case.
Because $\dim(G) = \dim(V)$ the the vectors $A_1 v, \ldots, A_n v$ are linearly independent for any element $v \in U$. Hence if $S_{ij}(x) = 0$ vanishes on U, then each $s_i(x)$ vanishes on U and because U is open and dense in V, each $s_i(x)$ vanishes on V. We conclude that $s_1 = \ldots = s_n = 0$ and hence $\tilde{\mathfrak{g}}$ is a free $\mathcal{O}_V$–submodule of $\mathrm{Der}(-log\ D)$. □

Lemma 4.1.2. We have $\tilde{\mathfrak{g}} = \mathcal{O}_V E \oplus \tilde{\mathfrak{a}}$.
In particular $\tilde{\mathfrak{a}}$ is a free $\mathcal{O}_V$–submodule of $\mathcal{D}er(-log\ h)$.

Proof. We recall that the Lie algebra $\mathfrak{g}$ of $G \leq \mathrm{GL}_n(\mathbb{C})$ is characterized by

$$\mathfrak{g} \cong \{A \in \mathrm{Mat}_{n \times n}(\mathbb{C}) \mid \exp(tA) \in G,\ \forall t \in \mathbb{C}\},$$

where $\exp(tA) := \sum_{i=0}^{\infty} \frac{t^n}{n!} A^n \in \mathrm{GL}_n(\mathbb{C})$.
We denote by $\mathbf{1}$ the identity matrix of $\mathrm{Mat}_{n \times n}(\mathbb{C})$ and claim that $\mathbf{1} \in \mathfrak{g}$. In this case $E = \sum_{i=1}^{n} x_i \partial_i$ is an element of $\tilde{\mathfrak{g}}$ and the formula

$$\xi = \frac{\xi(h)}{E(h)} E + (\xi - \frac{\xi(h)}{E(h)} E)$$

for $\xi \in \tilde{\mathfrak{g}}$ defines the claimed decomposition of $\tilde{\mathfrak{g}}$.
For $t \in \mathbb{C}$ we denote by $g_t = \exp(t\mathbf{1}) = \exp(t)\mathbf{1} \in \mathrm{GL}_n(\mathbb{C})$. Obviously

the claim $\mathbf{1} \in \mathfrak{g}$ is equivalent to the statement, that we have for all $t \in \mathbb{C}$ and for all $v \in D$

$$h(g_t.v) = 0.$$

Because h is homogeneous, we have $h(\lambda x) = \lambda^{n_h} h(x)$ for all $\lambda \in \mathbb{C}$. Hence we conclude that for all $v \in h^{-1}(0) = D$

$$h(g_t.v) = \exp(t)^{n_h} h(v) = 0$$

holds. Therefore $\exp(t\mathbf{1})$ lies in G for all t. □

We will now introduce the notions of $\mathcal{R}$–finiteness:

Definition 4.1.3. We set

$$\begin{aligned} T_{\mathcal{R}_D} f &= \frac{\mathcal{O}_V}{df(\mathrm{Der}(-\log\, D))} \\ T_{\mathcal{R}_h} f &= \frac{\mathcal{O}_V}{df(\mathrm{Der}(-\log\, h)) + (h)} \\ T_{\mathcal{R}_\mathfrak{g}} f &= \frac{\mathcal{O}_V}{df(\tilde{\mathfrak{g}})} \\ T_{\mathcal{R}_\mathfrak{a}} f &= \frac{\mathcal{O}_V}{df(\tilde{\mathfrak{a}}) + (h)} \end{aligned}$$

and say $f \in V^*$ is $\mathcal{R}_D$*–finite* if $\dim_\mathbb{C}(T_{\mathcal{R}_D} f) < \infty$ and in the same way $\mathcal{R}_h$–, $\mathcal{R}_\mathfrak{g}$– or $\mathcal{R}_\mathfrak{a}$–finite, if $\dim_\mathbb{C}(T_{\mathcal{R}_h} f)$, $\dim_\mathbb{C}(T_{\mathcal{R}_\mathfrak{g}} f)$ or $\dim_\mathbb{C}(T_{\mathcal{R}_\mathfrak{a}}) < \infty$.

Remark 4.1.4. We recall that the map $\phi\colon \mathfrak{g} \to \mathrm{Der}(-\log\, D)$ is injective, hence $\mathcal{R}_\mathfrak{g}$–finiteness implies $\mathcal{R}_D$–finiteness and $\mathcal{R}_\mathfrak{a}$–finiteness implies $\mathcal{R}_h$–finiteness. In particular by lemma 1.4.1 the induced map $\phi\colon \tilde{\mathfrak{g}} \to \mathrm{Der}(-\log\, D)$ is an isomorphism if and only if D is a linear free divisor and in this situation the notions of $\mathcal{R}_D$ and $\mathcal{R}_\mathfrak{g}$ (resp. $\mathcal{R}_h$ and $\mathcal{R}_\mathfrak{a}$) and the corresponding modules coincide.

Our next aim is to prove that if $f \in V^*$ is $\mathcal{R}_\mathfrak{a}$–finite, then f is $\mathcal{R}_\mathfrak{g}$–finite and the module $T_{\mathcal{R}_\mathfrak{a}} f = \frac{\mathcal{O}_V}{df(\tilde{\mathfrak{a}})+(h)}$ is generated by the powers $1, f^1, \ldots, f^{n_h - 1}$, where $n_h = \deg(h)$. But this needs some preparation:

Lemma 4.1.5. The G–orbit of f is open in V^* if and only if f is $\mathcal{R}_g$–finite.
In particular in this case $df(\tilde{\mathfrak{g}})$ and $d\alpha_f(\mathfrak{g})$ span the maximal ideal $\mathfrak{m}$ at zero of $\mathcal{O}_V$.

Proof. First we remark that the ideal $df(\tilde{\mathfrak{g}})$ is generated by linear functions, as f is linear. Because $\dim(G) = \dim(V)$ we have that f is $\mathcal{R}_\mathfrak{g}$–finite if and only if $d\alpha_f(\mathfrak{g})$ spans the maximal ideal $\mathfrak{m}$ at zero. This holds if and only if the differential of the orbit map $d\alpha_f$ is surjective, which is the case if and only if the orbit of f is open in V^*. □

We set $L_f := \mathcal{V}(df(\mathrm{Der}(-log\ h)))$ and $L_{\mathfrak{a}} := \mathcal{V}(df(\tilde{\mathfrak{a}}))$. Because $\tilde{\mathfrak{a}}$ is generated by vector fields of weight zero, the module $df(\tilde{\mathfrak{a}})$ is generated by linear forms and hence the space $L_{\mathfrak{a}}$ is a linear subspaces of V.

Lemma 4.1.6. The space $L_{\mathfrak{a}}$ is a line transversal to $\mathcal{V}(f)$ if and only if f is $\mathcal{R}_{\mathfrak{g}}$–finite.

Proof. Because the Euler field E lies in $\tilde{\mathfrak{g}}$, the claim follows from

$$df(\tilde{\mathfrak{g}}) = df(\tilde{\mathfrak{a}}) + (df(E)) = df(\tilde{\mathfrak{a}}) + (f).$$

Indeed as $L_{\mathfrak{a}}$ and $\mathcal{V}(f)$ are linear subspaces of V, they meet transversally if and only if $\dim(L_{\mathfrak{a}}) + \dim(\mathcal{V}(f)) = \dim(V)$. This holds if and only if $df(\tilde{\mathfrak{g}})$ generates the maximal ideal of $\mathcal{O}_V$ at zero, which holds by the proof of 4.1.5 if and only if the G–orbit of f is open in V^*. Now the claim follows from lemma 4.1.5. □

Lemma 4.1.7. If $L_{\mathfrak{a}} \subset V$ is a line, then we either have $L_{\mathfrak{a}} \cap D = \{0\}$ or $L_{\mathfrak{a}} \subset D$. In particular $L_{\mathfrak{a}} \cap D = \{0\}$ holds if and only if f is $\mathcal{R}_{\mathfrak{a}}$–finite. In particular in this case $d\alpha_f(A_1), \dots, d\alpha_f(A_{n-1})$ is a regular sequence in $\mathcal{O}_V$ for any basis $A_1, \dots, A_{n-1}$ of $\mathfrak{a}$.

Proof. Assume there is a $p \in L_{\mathfrak{a}} \cap D$ with $p \neq 0$. Because the defining equation h of D is homogeneous, we have for all $k \in \mathbb{C}$:

$$h(kp) = k^{\deg(h)} h(p) = 0,$$

i.e. we have $L_{\mathfrak{a}} \subset D$. As this is equivalent to $h \in df(\mathfrak{a})$, f is not $\mathcal{R}_{\mathfrak{a}}$–finite. The remaining statements are trivial. □

Lemma 4.1.8. (cf. [GMS09] Prop. 3.5 (ii)) If $f = 0$ is an equation for the tangent plane of $T_p D_t$, then

$$H(p) \neq 0 \Rightarrow \mu(f_{|D_t}, p) = 1,$$

where H denotes the Hessian determinant of h and $D_t = h^{-1}(t)$ for $t \in \mathbb{C}^*$.

Proof. The tangent space at p is $T_p D_t = \{v \mid \sum v_i \partial_i h(p) = 0\}$. So if f is a defining equation for $T_p D_t$, then $d_p h$ is a scalar multiple of $f = d_p f$. Let $\phi \colon \mathbb{C}^{n-1} \to D_t$ be a parametrisation of D_t around p. Obviously $h \circ \phi$ is constant. Because f is linear, we have

$$\frac{\partial^2 (f \circ \phi)}{\partial_{u_i} \partial_{u_j}} = \sum_s \frac{\partial f}{\partial_{x_s}} \frac{\partial^2 \phi_d}{\partial_{u_i} \partial_{u_j}}$$

and because $h \circ \phi$ is constant, we have

$$0 = \sum_{s,t} \frac{\partial^2 h}{\partial_{x_s} \partial_{x_t}} \frac{\partial \phi_s \partial \phi_t}{\partial_{u_i} \partial_{u_j}} + \sum_s \frac{\partial h}{\partial_{x_s}} \frac{\partial^2 \phi_s}{\partial_{u_i} \partial_{u_j}}.$$

So up to a non-zero scalar multiple we have the equality of $n \times n-$matrices:

$$(\frac{\partial^2 (f \circ \phi)}{\partial_{u_i} \partial_{u_j}}) = (\frac{\partial \phi_s}{\partial_{u_i}})^T (\frac{\partial^2 h}{\partial_{x_s} \partial_{x_t}} \circ \phi)(\frac{\partial \phi_t}{\partial_{u_j}}).$$

Hence if $H(p) \neq 0$, then the restriction $f_{|D_t}$ has a non-degenerated critical point at p. □

Lemma 4.1.9. For all $p \in L_{\mathfrak{a}}$ holds : ker $d_p f \subset \ker d_p h$.

Proof. If $p \in L_{\mathfrak{a}}$ is a singular point of h, then the inclusion is true as $\ker d_p h = V$. Hence let $p \in L_{\mathfrak{a}}$ be a regular point of h. In this case the hypersurfaces ker $d_p f$ and ker $d_p h$ are either equal or meet transversally. The equality holds if and only if

$$\text{rank}(\begin{pmatrix} \partial_1 f(p) & \dots & \partial_n f(p) \\ \partial_1 h(p) & \dots & \partial_n h(p) \end{pmatrix}) = 1.$$

For $\xi = \sum_i \xi_i(x) \partial_i \in \tilde{\mathfrak{a}}$ we have $\xi h = 0$ and in particular we have for all $p \in L_{\mathfrak{a}}$:

$$(\xi h)(p) = \sum_i \xi_i(p) \partial_i h(p) = 0.$$

On the other hand p lies in $L_{\mathfrak{a}}$ if and only if for all $\xi \in \tilde{\mathfrak{a}}$ holds

$$\sum_i \xi_i(p) \partial_i f(p) = 0.$$

Hence for all $\xi \in \tilde{\mathfrak{a}}$ holds :

$$\begin{pmatrix} \partial_1 f(p) & \dots & \partial_n f(p) \\ \partial_1 h(p) & \dots & \partial_n h(p) \end{pmatrix} \xi(p) = 0.$$

As $\mathfrak{a}$ is free of rank $n-1$ and $p \in \text{Reg}(h)$, we have

$$\dim(\ker \begin{pmatrix} \partial_1 f(p) & \dots & \partial_n f(p) \\ \partial_1 h(p) & \dots & \partial_n h(p) \end{pmatrix}) = n - 1,$$

and

$$\text{rank}(\begin{pmatrix} \partial_1 f(p) & \dots & \partial_n f(p) \\ \partial_1 h(p) & \dots & \partial_n h(p) \end{pmatrix}) = 1.$$

We conclude that $\ker(d_p f) = \ker(d_p h)$ holds for all regular points of h. □

Remark 4.1.10. We remark that the statement $\ker d_p f \subset \ker d_p h$ also holds for all $p \in L_f = \mathcal{V}(df(\mathrm{Der}(-log\ h))$. Indeed, the regularity of h at p implies

$$\dim(\ker \begin{pmatrix} \partial_1 f(p) & \dots & \partial_n f(p) \\ \partial_1 h(p) & \dots & \partial_n h(p) \end{pmatrix}) \leq n-1$$

and the existence of the $n-1$ dimensional free subalgebra $\mathfrak{a}$ in $\mathrm{Der}(-log\ h)$ implies

$$\dim(\ker \begin{pmatrix} \partial_1 f(p) & \dots & \partial_n f(p) \\ \partial_1 h(p) & \dots & \partial_n h(p) \end{pmatrix}) \geq n-1.$$

Proposition 4.1.11. (cf. [GMS09] Proposition 3.5 (iii)) If f is $\mathcal{R}_\mathfrak{a}$–finite, then

1. f is $\mathcal{R}_\mathfrak{g}$–finite,
2. the classes $1, f, \dots, f^{n_h-1}$ form a $\mathbb{C}$–basis of $T^1_{\mathcal{R}_\mathfrak{a}} = \frac{\mathcal{O}_{V,0}}{df(\tilde{\mathfrak{a}})+(h)}$,
3. on each Milnor fibre $D_t := h^{-1}(t)$ for $t \neq 0$, f has n_h non-degenerated critical points, which form an orbit under the diagonal action of μ_{n_h}, the group of n_h–th roots of unity.

Proof. 1. Let $A_1, \dots, A_{n-1}$ be a basis of $\mathfrak{a}$, which we identify with the free generators $\phi(A_1), \dots, \phi(A_{n-1})$ of the submodule $\tilde{\mathfrak{a}}$ of $\mathrm{Der}(-log\ h)$. If f is $\mathcal{R}_\mathfrak{a}$–finite, i.e. $\dim_\mathbb{C}(\frac{\mathcal{O}_V}{df(\tilde{\mathfrak{a}})+(h)}) < \infty$, then $df(A_1), \dots, df(A_{n-1})$ is a regular sequence in $\mathcal{O}_V$. Hence $L_\mathfrak{a} \subset V$ is a line and $L_\mathfrak{a} \cap D = \{0\}$ by lemma 4.1.7.
If f is not $\mathcal{R}_\mathfrak{g}$–finite, then f is constant on $L_\mathfrak{a}$. We conclude from lemma 4.1.9 that h is constant on $L_\mathfrak{a}$, which contradicts the assumption of f being $\mathcal{R}_\mathfrak{a}$–finite, i.e. $\dim_\mathbb{C}(\frac{\mathcal{O}_V}{df(\tilde{\mathfrak{a}})+(h)}) < \infty$.

Indeed let $p \in L_\mathfrak{a} \backslash \{0\}$. Then h is constant on the line $L_\mathfrak{a}$ if and only if we have for all $k \in \mathbb{C}$

$$0 = \partial_k h(kp) = \sum_{i=1}^{n} \frac{\partial h}{\partial_{x_i}}(kp) p_i,$$

but this holds by Lemma 4.1.9, if $p \in \ker f = \ker d_p f \subset \ker d_p h$.

2. We have seen that $L_\mathfrak{a}$ is a line, which intersects D only in 0. Moreover the class $\overline{h} \in \mathcal{O}_{L_\mathfrak{a}}$ is the n_h-th power of a generator of the maximal ideal $\mathfrak{m}_{L_\mathfrak{a}}$ at zero in $\mathcal{O}_{L_\mathfrak{a}}$.
In particular every $\tilde{f} \in \mathbb{C}[V]_{=1}$ with $\tilde{f} \notin df(\tilde{\mathfrak{a}})$ fulfils the condition

$$(\tilde{f}) + df(\tilde{\mathfrak{a}}) = \mathfrak{m},$$

where $\mathfrak{m} \subset \mathcal{O}_V$ denotes the maximal ideal at zero. We conclude that $\frac{\mathcal{O}_{L_\mathfrak{a}}}{(h)}$ is generated by $1, \tilde{f}, \ldots, \tilde{f}^{n_h-1}$.

In particular f itself fulfils this condition. Indeed because f is $\mathcal{R}_\mathfrak{g}$–finite, we have $df(\tilde{\mathfrak{a}}) + (f) = \mathfrak{m}$ and $f \notin df(\tilde{\mathfrak{a}})$.

3. By 1. f is $\mathcal{R}_\mathfrak{g}$–finite, so $L_\mathfrak{a}$ intersects $\mathcal{V}(f)$ transversally. The critical points of f_{D_t} are $\{p \in D_t \mid T_pD_t = \{f = 0\}\}$, hence $L_\mathfrak{a}$ intersects D_t transversally in each critical point. The ideals $df(\tilde{\mathfrak{a}})$ and $J_{f_{|D_t}}$ coincide in $\mathcal{O}_{D_t}$, hence the intersection number, which is equal to 1, equals the Milnor number of $f_{|D_t}$ at p.
   That there are n_h critical points, counting multiplicity, follows from the fundamental theorem of algebra, applied to the single variable polynomial $(h-t)_{L_\mathfrak{a}}$. Because h and $L_\mathfrak{a}$ are μ_{n_h} invariant, these points form an orbit under the diagonal action of μ_{n_h}.

□

We derive from the proof of proposition 4.1.11:

Corollary 4.1.12. If $f \in V^*$ is $\mathcal{R}_\mathfrak{a}$–finite, then we have

$$\frac{\mathcal{O}_V}{df(\tilde{\mathfrak{a}})} \cong \mathbb{C}[f].$$

Proof. We denote by $\mathfrak{m}$ the maximal ideal at zero of $\mathcal{O}_V$. Because $f \in V^*$ is $\mathcal{R}_\mathfrak{a}$–finite we have

$$(f) + df(\tilde{\mathfrak{a}}) = \mathfrak{m}.$$

We conclude that

$$\frac{\mathfrak{m}}{df(\tilde{\mathfrak{a}}) + \mathfrak{m}^2} = \langle \overline{f} \rangle_\mathbb{C}$$

and for all $k \in \mathbb{N} \backslash \{0\}$

$$\frac{\mathfrak{m}^k + df(\tilde{\mathfrak{a}})}{df(\tilde{\mathfrak{a}}) + \mathfrak{m}^{k+1}} = \langle \overline{f}^k \rangle_\mathbb{C}.$$

Hence we have

$$df(\tilde{\mathfrak{a}}) + \mathbb{C}[f] = \mathcal{O}_V.$$

□

4.2 $\mathcal{R}$–finiteness for reductive groups

We use the notation from the introduction of this chapter and recall from the previous section that $\mathcal{R}_{\mathfrak{a}}$–finiteness implies $\mathcal{R}_{\mathfrak{g}}$–finiteness.
In this section we prove that the notion of $\mathcal{R}_{\mathfrak{g}}$–finiteness and $\mathcal{R}_{\mathfrak{a}}$–finiteness coincide, when G is a reductive group.

Lemma 4.2.1. (cf. [GMS09] Lemma.3.9) Let $D \subset V$ be the discriminant of a regular reductive vector space. Let h be an equation for D of degree n_h and denote by H the Hessian determinant of h. Then we have

$$h^*(\partial_1 h, \ldots, \partial_n h) = \frac{1}{n_h - 1} H\tilde{h},$$

where $\tilde{h} = \det(\phi(A_1)x| \ldots |\phi(A)_n x)$ and $A_1, \ldots, A_n$ is a basis of $\mathfrak{g}$.

Proof. We choose a basis $A_1 = \mathbf{1}, A_2, \ldots, A_n$ of $\mathfrak{g}$, such that $A_2, \ldots, A_n \in \mathfrak{a}$. We write $A_i = [a_{ij}^k]$, where the upper index refers to the columns and the lower index refers to the rows of A_i.
We denote by $\alpha_{ij} = \sum_k a_{ij}^k x_k$ the coefficient of ∂_j of the vector field $\xi_i = x^T A_i^T \partial$. Then we have for $i = 2, \ldots, n$:

$$0 = \xi_i(h) = \sum_j \alpha_{ji} \partial_j(h).$$

We differentiate the equations with respect to x_k :

$$0 = \sum_j \frac{\partial \alpha_{ji}}{\partial_k} \frac{\partial h}{\partial_j} + \sum_j \alpha_{ji} \frac{\partial^2 h}{\partial_k \partial_j} = \sum_j a^k ij \frac{\partial h}{\partial_j} + \sum_j \alpha_{ji} \frac{\partial^2 h}{\partial_k \partial_j}. \tag{4.1}$$

For $\xi_1(h) = n_h \cdot h$ we have

$$n_h \cdot h = \xi_1(h) = \sum_j \alpha_{j1} \partial_j h$$

and therefore

$$n_h \cdot \partial_k h = \sum_j a_{1,j}^k \partial_j h + \sum_j \alpha_{j1} \frac{\partial^2 h}{\partial_k \partial_j} = \partial_k(h) + \sum_j \alpha_{j1} \frac{\partial^2 h}{\partial_k \partial_j}. \tag{4.2}$$

Hence writing the equations (4.1) and (4.2) in matrix form, we get

$$\begin{pmatrix} E^T \\ \xi_2^T \\ \vdots \\ \xi_n^T \end{pmatrix} \left(\frac{\partial^2 h}{\partial_k \partial_j} \right) = - \begin{pmatrix} (n_h - 1) \nabla h \cdot \mathbf{1} \\ \nabla h \cdot A_2 \\ \vdots \\ \nabla h \cdot A_n \end{pmatrix}.$$

We now take determinants of both sides. The determinant on the right hand side is

$$(n_h - 1)h^*(\partial_1 h, \ldots, \partial_n h)$$

and the determinants of the left hand side are $\tilde{h} = \det(\phi(A_1)x| \ldots |\phi(A)_n x)$ and H. $\square$

Lemma 4.2.2. ([SK77] p.71-72) If $D \subset V$ is the discriminant of a regular reductive vector space with equation h of degree n_h, then for all $p \in V$ holds:

$$h(p) \neq 0 \Rightarrow H(p) \neq 0.$$

Proof. We recall from section 1.3 that if g is any homogeneous semi-invariant in a reductive prehomogeneous vector space (G, V) of degree n_g with associated character χ_h, then there exists a polynomial $b_g(s)$, such that we have in unitary coordinates on V :

$$g^*(\partial_1, \ldots, \partial_n)g^s = b_g(s)g^{s-1}.$$

Moreover $b(s)$ has precisely degree n_g and by [SK77] p.72 we have

$$g^*(\partial_1 g, \ldots, \partial_n g)g^{s-n_g} = b_0 g^{s-1}, \tag{4.3}$$

where b_0 denotes the leading coefficient of $b_g(s)$.
Now we specialize to the situation, where the semi-invariant $g = h$ is an equation for the discriminant D, i.e. we have by equation (4.3)

$$h^*(\partial_1 h, \ldots, \partial_n h)h^{s-n_h} = b_0 h^{s-1}. \tag{4.4}$$

We conclude from equation (4.4) that $h^*(\partial_1 h, \ldots, \partial_n h) = b_0 h^{n_h-1}$. We apply Lemma 4.2.1 and get

$$H = (n-1)b_0 \frac{h^{n_h-1}}{\tilde{h}}.$$

Now h and $\tilde{h}$ are both equations for D, i.e. we have $h^{-1}(0) = \tilde{h}^{-1}(0) = D$. Hence the claim of this lemma follows from lemma 4.2.1. $\square$

Lemma 4.2.3. (cf. [GMS09] Lemma. 3.11) Let h be an equation of the discriminant of a reductive prehomogeneous vector space (G, V) with respect to unitary coordinates, then

1. the gradient ∇h maps the Milnor fibres of h diffeomorphically to the fibres of h^*,

2. the gradient ∇h^* maps the Milnor fibres of h^* diffeomorphically to the fibres of h.

Proof. By the formula $h^*(\partial_1 h, \ldots, \partial_h) = b_0 h^{n_h-1}$ from the proof of lemma 4.2.2, we know that ∇h maps fibres of h to fibres of h^*. We put $A = \ker(\chi_h : G \to \mathbb{C}^*)^0$ and $D_t = h^{-1}(t)$ for $t \neq 0$. Furthermore because G is reductive, we have $\chi_{h^*} = \chi_h^{-1}$ (cf. section 1.3). Hence we have $A = \ker \chi_{h^*}$.

We note that D_t is an A orbit for $t \neq 0$. Indeed for $a \in A$ and $x \in D_t$ we have $h(ax) = \chi_h(a)h(x) = 1 \cdot t$. Hence we have $A.D_t \subset D_t$. On the other hand for all $x, y \in D_t$ is a $g \in G$, such that $y = g.x$. We conclude from $t = h(y) = h(g.x) = \chi(g) \cdot t$ that g lies in A. Moreover the map ∇h is equivariant, because of

$$\nabla h(\rho(g)x) = \rho^*(g)^{-1} \nabla h(x).$$

Hence it is surjective and by Lemma 4.2.2 it is a local diffeomorphism. The second claim follows from the first by interchanging h and h^* in the arguments. This works, because $(h^*)^* = h$ and the dual coordinates of the unitary coordinates are unitary coordinates themselves. □

Theorem 4.2.4. Let D be the discriminant of a reductive prehomogeneous vector space (G, ρ, V), then $f \in V^*$ is $\mathcal{R}_\mathfrak{a}$–finite if and only if f is $\mathcal{R}_\mathfrak{g}$–finite.

Proof. Because we have shown one direction in proposition 4.1.11, we only have to show: If f is $\mathcal{R}_\mathfrak{g}$–finite, then f is $\mathcal{R}_\mathfrak{a}$–finite.

We recall from lemma 4.1.7 that f is $\mathcal{R}_\mathfrak{a}$–finite if and only if $L_\mathfrak{a} = \mathcal{V}(df(\tilde{\mathfrak{a}}))$ is a line in V and $L_\mathfrak{a} \cap D = \{0\}$.

Let f be $\mathcal{R}_\mathfrak{g}$ finite. Then $L_\mathfrak{a}$ is a line by lemma 4.1.6. Hence it remains to prove that $L_\mathfrak{a} \cap D = \{0\}$.

Because f is $\mathcal{R}_\mathfrak{g}$–finite, we have $f \in V^* \setminus D^*$ by lemma 4.1.5. Hence by lemma 4.2.3 f is a non-zero multiple of $\nabla h(p)$ for some $p \notin D$, i.e. $\mathcal{V}(f) = T_p D_t$ for $t \neq 0$.

Because we have $h(p) \neq 0$, we conclude that from lemma 4.2.2 that $f_{|D_t}$ has a non-degenerate critical point at p and from lemma 4.1.7 we conclude that $L_\mathfrak{a}$ intersects D_t transversal. As $L_\mathfrak{a}$ is a linear subspace of V we write $L_\mathfrak{a} = \{\lambda q | \lambda \in \mathbb{C}\}$ for a $q \in L_\mathfrak{a} \cap D_t$. Now a point λq of $L_\mathfrak{a}$ lies in $L_\mathfrak{a} \cap D$ if $h(\lambda q) = \lambda^n t = 0$. Of course this holds only for $\lambda = 0$ as $t \neq 0$. We conclude that $L_\mathfrak{a} \cap D = \{0\}$ and hence f is $\mathcal{R}_\mathfrak{a}$–finite by lemma 4.1.7. □

4.3 Restrictions to the Milnor fiber

Let (G, ρ, V) be a regular reductive prehomogeneous vector space and h an equation for the discriminant D.
We know from corollary 4.1.12, that for $f \in V^* \backslash D^*$ we have:

$$\frac{\mathcal{O}_V}{df(\tilde{\mathfrak{a}})} \cong \mathbb{C}[f].$$

Furthermore we have by proposition 3.2.6:

$$\frac{\Omega_{rel}^{n-1}(log\ D)}{df \wedge \Omega_{rel}^{n-2}(log\ D)} \cong \frac{\mathcal{O}_V}{df(\mathrm{Der}(-log\ h))}[\alpha].$$

We want to know, if this modul is generated by the powers of f. Therefore we need to know, when the inclusion

$$df(\tilde{\mathfrak{a}}) \subset df(\mathrm{Der}(-log\ h))$$

is an equality.

At first we remark that the left hand side defines a line through the origin. Indeed as f lies in the open orbit, we conclude from lemma 4.1.5 that f is $\mathcal{R}_{\mathfrak{g}}$–finite. As the group G is reductive, this is equivalent by theorem 4.2.4 to f is $\mathcal{R}_{\mathfrak{a}}$–finite, which is by lemma 4.1.7 equivalent to the statement that $df(\tilde{\mathfrak{a}})$ defines a line though the origin.

At second we remark that the right hand side either defines the same line, if the ideals $df(\tilde{\mathfrak{a}})$ and $df(\mathrm{Der}(-log\ h)$ coincide or, if the inequality is strict, defines the origin with a non reduced structure. Indeed, because all derivations of degree zero lie in $\phi(\mathfrak{g})$ by lemma 1.4.1 the set theoretic difference $\mathrm{Der}(-log\ h)\backslash\tilde{\mathfrak{a}}$ consists of derivations of degree bigger than 1. Hence the ideal $df(\mathrm{Der}(-log\ h))$ contains a power $\mathfrak{m}^k$, for $k > 1$, of the maximal ideal $\mathfrak{m}$ at zero.

If $df(\mathrm{Der}(-log\ h))$ defines a non reduced structure at the origin, then the restriction $f_{|D_t}$ to the Milnor fibres $D_t = h^{-1}(t)$, for $t \neq 0$, is non singular. Indeed the critical locus of $f_{|D_t} : D_t \to \mathbb{C}$ is $L_f = \mathcal{V}(\mathrm{Der}(-log\ h))$ and by assumption we have $L_f \subset D$.

On the other hand as f is $\mathcal{R}_{\mathfrak{g}}$–finite the line $L_{\mathfrak{a}}$ intersects D_t transversally in some point p and by the proof of theorem 4.2.4 the restriction $f_{|D_t}$ has a non-degenerated critical point at p.

We summarize the discussion:

Proposition 4.3.1. Let (G, ρ, V) be a regular reductive prehomogeneous vector space and h be an equation for the discriminant. Then for all $f \in V^* \backslash D^*$ the restriction $f_{|D_t} \colon D_t \to \mathbb{C}$ is singular and

$$df(\tilde{\mathfrak{a}}) = df(\mathrm{Der}(-log\ h)).$$

Furthermore by the homogeneity of the equation h the line $L_{\mathfrak{a}}$ intersects each Milnor fibre D_t.

Chapter 5

The Gauss-Manin System associated to a linear section of the Milnor fibration

Let $\Phi\colon X \to Y$ be a morphism of smooth complex algebraic varieties. In $\mathcal{D}-$module theory the *Gauss-Manin system* is defined as the derived direct image $\Phi_+\mathcal{O}_X$ of the structure sheaf $\mathcal{O}_X$, seen as a left $\mathcal{D}_X-$module, i.e.

$$\Phi_+\mathcal{O}_X := \mathbf{R}\Phi_*(\mathcal{D}_{Y\leftarrow X} \otimes^{\mathbf{L}}_{\mathcal{D}_X} \mathcal{O}_X).$$

In the case that X is affine of dimension n and $\Phi\colon X \to \mathbb{A}^1$ is a regular function, then by [Sab06] §9 the Gauss-Manin system $\Phi_+\mathcal{O}_X$ is represented by the shifted complex

$$(\Phi_*\Omega_X^{n+\bullet}[\partial_t], d_\Phi),$$

where d_Φ is defined by $d_\phi(\sum_i \omega_i\partial_t^i) = \sum_i d\omega_i\partial_t^i - \sum_i d\Phi \wedge \omega_i\partial_t^{i+1}$ and t is a coordinate on $\mathbb{A}^1$. The cohomology modules $\mathcal{H}^k$ are naturally equipped with a $\mathbb{C}[t]\langle\partial_t\rangle-$ structure. In particular the $\mathcal{H}^k$ are holonomic and $\mathcal{H}^j(\Phi_+\mathcal{O}_X) = 0$ holds for $j \neq [-n, 0]$.
The $\mathbb{C}[t]\langle\partial_t\rangle-$structure is given by

$$\begin{aligned} \partial_t.(\omega \otimes P) &= \omega \otimes P\partial_t \\ g(t).(\omega \otimes P) &= g(t)\omega \otimes P, \end{aligned}$$

for $P \in \mathbb{C}[\partial_t]$.

To give some references for the theory on $\mathcal{D}-$modules we name [Bor87] and [HT07] for algebraic $\mathcal{D}-$modules and [Bjö93] for analytical $\mathcal{D}-$modules.

5.1 The localized Gauss-Manin system

Let $V = \mathbb{A}^n_{\mathbb{C}}$ and $h\colon V \to T = \mathrm{Spec}(\mathbb{C}[t])$, $f\colon V \to R = \mathrm{Spec}(\mathbb{C}[r])$ be morphisms of affine varieties. We denote by $D = \mathcal{V}(h)$ the zero set of h and assume that h is given by a homogeneous polynomial. We consider the morphism $\Phi := (f,h)\colon V \to R \times T$, but instead of studying the direct image of the structure sheaf, we will consider the direct image of $\mathcal{O}_V(*D)$, the sheaf of rational functions on V with poles along the divisor D.
We will see that in this situation the direct image will be given by a complex of modules of relative differential forms and we will derive a formula for the ∂_t- and ∂_r-action in this particular situation.

Definition 5.1.1. The *localized Gauss-Manin system of the pair* (f,h) is the direct image complex $\Phi_+\mathcal{O}_V(*D)$ in $D^b(\mathcal{D}_{R\times T})$.
As usual we compute the direct image complex by decomposing Φ into the graph embedding $\Gamma_\Phi\colon V \to V \times R \times T$, $v \mapsto (v, f(v), h(v))$ and the projection $\pi\colon V \times R \times T \to R \times T$, $(v,r,t) \mapsto (r,t)$, as these two cases can be handled explicitly. Indeed because we are working in the algebraic category, we have by [HT07] prop.1.5.21 : $\Phi_+ = \pi_+ \circ (\Gamma_\Phi)_+$.
We refer to [Bor87] V §3 or [HT07] for the details of this calculation, but as in the case of a regular function, we get for coordinates with associated partial derivatives $(x_1,\dots,x_n,\partial_1,\dots,\partial_n)$ on V and (r,∂_r) on R and (t,∂_t) on T a representation of the direct image complex by

$$\Phi_+\mathcal{O}_V(*\,D) = \mathbf{R}\Phi_*(\Omega_V^{n+\bullet}(*\,D)[\partial_r,\partial_t],\nabla)$$

where ∇ is defined as

$$\nabla(\omega\otimes P) = d\omega\otimes P - df\wedge\omega\otimes P\partial_r - dh\wedge\omega\otimes P\partial_t$$

and the $\mathcal{D}_{R\times T}-$structure on the cohomology is given by

$$\begin{aligned} \partial_r.(\omega\otimes P) &= \omega\otimes P\partial_r && (5.1)\\ \partial_t.(\omega\otimes P) &= \omega\otimes P\partial_t && (5.2)\\ g(r,t).(\omega\otimes P) &= g(r,t)\omega\otimes P, \end{aligned}$$

for $P \in \mathbb{C}[\partial_r,\partial_t]$.
As Φ is affine, the direct image functor Φ_* is exact and we have

$$\Phi_+\mathcal{O}_V(*\,D) = \Phi_*(\Omega_V^{n+\bullet}(*\,D)[\partial_r,\partial_t],\nabla).$$

Remark 5.1.2. Let R be a ring and $X = \mathrm{Spec}(R)$. Without mentioned it we have used the well known equivalence of categories between the category of quasi coherent sheaves of $\mathcal{O}_X-$modules and the category of $R-$modules given by the global section functor. Similar is true for $\mathcal{D}-$modules. Precisely; Let X be a smooth affine algebraic variety. By [HT07] 1.4 there is an equivalence between the category of quasi coherent sheaves of $\mathcal{D}_X-$modules and the category of $\Gamma(X, \mathcal{D}_X)-$modules. Furthermore as for $\mathcal{O}_X-$modules this equivalence induces an equivalence

$$\mathrm{Mod}_c(\mathcal{D}_X) \simeq \mathrm{Mod}_{f.g.}(\Gamma(X, \mathcal{D}_X)),$$

where $\mathrm{Mod}_c(\mathcal{D}_X)$ denotes the category of coherent $\mathcal{D}_X-$modules and $\mathrm{Mod}_{f.g.}(R)$ denotes the category of finitely generated $R-$modules.
We remark that this equivalence does not only hold for affine algebraic varieties. For example by [HT07] theorem 1.6.5 it is true for projective spaces. More precisely, this equivalence holds for any $D-$affine variety X.We refer for the definition and the details to [HT07] 1.4.

Remark 5.1.3. We remark that we have defined the localized Gauss-Manin system as the direct image of sheaf $\mathcal{O}_V(*D)$ of rational functions along D, but we get the same result if we first apply the direct image functor to the structure sheaf $\mathcal{O}_V$ and localize afterwards along $t = 0$. Because of the localization we have the following:

Lemma 5.1.4. The complex $(\Omega_V^\bullet(*D), dh\wedge)$ is acyclic.

Proof. Let R be a ring and M be a $R-$module. We recall the following property of the Koszul complex $K(x_1, \ldots, x_n)$ of $x_1, \ldots, x_n \in R$, see for example [Eis13] Prop. 17.14 (b):

If $(x_1, \ldots, x_n)M = M$, then $H^i(M \otimes_R K(x_1, \ldots, x_n)) = 0$ for all i.

We consider the situation where the ring R is $M = \mathcal{O}_V(*D)$ and the ideal is given by the partial derivatives of h, i.e. $(x_1, \ldots, x_n) = (\partial_1 h, \ldots, \partial_n h)$. It is well known that as complexes of $\mathcal{O}_V(*D)-$modules we have

$$(\Omega_V^\bullet(*\, D), dh\wedge) \cong K(\partial_1 h, \ldots, \partial_n h).$$

Because h is homogeneous, we have $h \in J_h = (\partial_1 h, \ldots, \partial_n h)$ and as h is invertible in $R = \mathcal{O}_V(*D)$ the Jacobian ideal J_h is equal to $\mathcal{O}_V(*D) = R$. So we conclude from the proposition that the Koszul complex $(\Omega_V^\bullet(*D), dh\wedge)$ of $\partial_1 h, \ldots, \partial_n h$ is acyclic. □

Lemma 5.1.5. The map

$$\begin{aligned} \phi^\bullet \colon (\Omega_{V/T}^{n-1+\bullet}(*D), d - df\wedge) &\rightarrow (\ker^{n+\bullet} dh\wedge, \delta) \\ \omega &\mapsto dh \wedge \omega \end{aligned}$$

is an isomorphism of complexes for $\delta = (-1) \cdot (d - df\wedge)$.

Proof. Let us first recall the definitions: $\Omega^\bullet_{V/T} = \Omega^\bullet_{rel} = \Omega^\bullet_h = \frac{\Omega^\bullet_V}{dh\wedge\Omega^{\bullet-1}_V}$ and $\ker^k dh\wedge := \ker dh\wedge\colon \Omega^k_V \to \Omega^{k+1}_V$.
That $\phi^\bullet$ is a map of complexes, i.e. $\delta \circ \phi^k = \phi^{k+1} \circ (d - df\wedge)$, follows from a direct calculation. So it remains to prove ϕ is an isomorphism. Let $[\omega_{k-1}] \in \Omega^{n+k-1}_{V/T}(*D)$ an element of the kernel of ϕ, i.e. $\omega_{k-1} \in \ker^{n+k-1} dh\wedge \subset \Omega^{n+k-1}_V(*D)$. Because $(\Omega^{n+\bullet}_V(*D), dh\wedge)$ is acyclic, there is a $\omega_{k-2} \in \Omega^{n+k-2}_V(*D)$, such that $dh \wedge \omega_{k-2} = \omega_{k-1}$, so $[\omega_{k-1}] = [0] \in \Omega^{n+k-1}_{V/T}$. Hence ϕ is injective.

If $\omega_k \in \ker^{n+k} dh$, then, because $(\Omega^{n+\bullet}_V(*D), dh\wedge)$ is acyclic, there is a $\omega_{k-1} \in \Omega^{n+k-1}_V(*D)$, such that $dh \wedge \omega_{k-1} = \omega_k$, i.e. ϕ maps the class $[\omega_{k-1}] \in \Omega^{n+k-1}_{V/T}$ to ω_k. Hence ϕ is surjective. $\square$

For $\omega \in \ker^{n+k} dh\wedge$ we denote by $[\frac{\omega}{dh}]$ its inverse in $\Omega^{n-1+k}_{V/T}(*D)$. Because $(\Omega_V(*D), dh\wedge)$ is acyclic, there is a $\omega_{k-1} \in \Omega^{n+k-1}_V(*D)$ such that $\omega = dh \wedge \omega_{k-1}$ and we conclude from the proof above $[\frac{\omega}{dh}] = [\omega_{k-1}]$.

Proposition 5.1.6. The complexes

$$\Omega^{n+\bullet}_V(*D)[\partial_r, \partial_t], d - df \wedge \_\partial_r - dh \wedge \_\partial_t) \simeq (\Omega^{n-1+\bullet}_{V/T}(*D)[\partial_r], d - df \wedge \_\partial_r)$$

are quasi isomorphic.

Proof. We prove the statement by a spectral sequence argument. We put $\Omega^{n+\bullet} := \Omega^{n+\bullet}_V(*D)[\partial_r]$, $D_f := d - df \wedge \_\partial_r$ and $D_h := -dh \wedge \_\partial_t$ and consider the double complex

$$(K^{p,q}, D^{p,q}_{hor}, D^{p,q}_{vert}) := (\Omega^{n+p+q}\partial_t^q, D_f, (-1)^q D_h).$$

Its associated total complex is equal to $(\Omega^{n+\bullet}_V(*D)[\partial_r, \partial_t], D_f + D_h)$. The zero page of the vertical spectral sequence lies in the upper half plane. Precisely, the picture looks like this:[1]

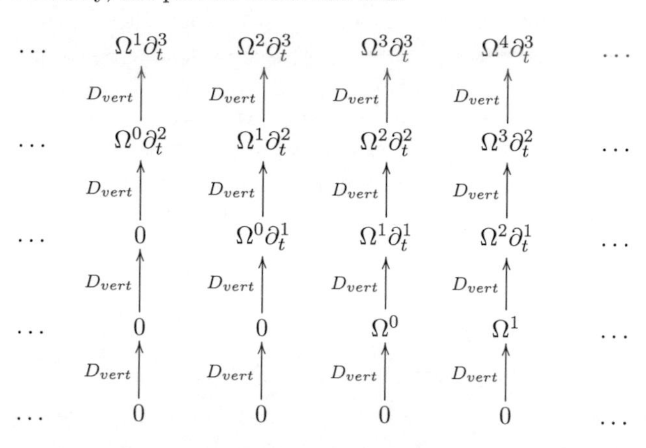

By lemma 5.1.4 non trivial cohomology groups can sit only in the bidegrees $(-k, 0)$ for $k \in \mathbb{Z}$. Hence the first page of the spectral sequence is given by the single complex

$$0 \to \ker^{n-(n-1)} D_h \to \ker^{n-(n-2)} D_h \to \dots$$

where $\ker^{n+k} D_h = \ker(D_h \colon \Omega^{n+k} \to \Omega^{n+k+1}\partial_t)$. This complex is by lemma 5.1.5 isomorphic to $(\Omega_{V/T}^{n-1+\bullet}[\partial_r](*D), d - df \wedge \_\partial_r))$. As the spectral sequence degenerates this complex has the same cohomology as the total complex. $\square$

Remark 5.1.7. We conclude that the direct image is represented by the complex:

$$\Phi_+\mathcal{O}_V(*D) = (\Phi_*\Omega_{V/T}^{n-1+\bullet}(*D)[\partial_r], d - df \wedge \_\partial_r).$$

So far we know, how we can represent the direct image $\Phi_+\mathcal{O}_V(*D)$ as a complex of $\mathcal{O}-$modules and it remains to understand the $\mathcal{D}-$module structure. For simplicity we write

$$M_k := \mathrm{H}^k(\Phi_*\Omega_{V/T}^{n-1+\bullet}(*D)[\partial_r], d - df \wedge \_\partial_r).$$

The ∂_r-action on M_k is given by multiplication from the right by formula (5.1). So it remains to understand the ∂_t-action.
Therefore we denote by

$$N_k := \mathrm{H}^k(\Phi_*\Omega_V^{n+\bullet}(*D)[\partial_r, \partial_t], d - df \wedge \_\partial_r - dh \wedge \_\partial_t),$$

[1] We remark that the module Ω^0 in the bottom row sits in bidegree $(-n, 0)$.

and by $[\cdot]_{N_k}$ a class in N_k and by $[\cdot]_{M_k}$ a class in M_k .
Notice that we only need to compute $\partial_t.[\omega_k]_{M_k}$ where $\omega_k \in \Phi_*\Omega^{n-1+k}_{V/T}(*D)$, since for a general element $[\sum_{i=0}^m \eta_i \partial_r^i]_{M_k}$ the derivative ∂_t can simply defined by linearity in ∂_r. Hence we let $\omega_k \in \Omega_V^{n-1+k}(*D)$ be a representative for a class $[\omega_k]_{M_k}$, in particular, we have $d\omega_k - df \wedge \omega_k \partial_r = 0$ in $\Phi_*\Omega^{n+k}_{V/T}(*D)$, which means that the two identities $dh \wedge d\omega_k = 0$ and $dh \wedge df \wedge \omega_k = 0$ hold true. According to Proposition 5.1.6, the class $[\omega_k]_{M_k}$ corresponds to the class $[dh \wedge \omega_k]_{N_k}$ under the isomorphism $M_k \cong N_k$. From formula (5.2) we conclude:

$$\partial_t.[dh \wedge \omega_k]_{N_k} = [dh \wedge \omega_k \partial_t]_{N_k}.$$

and from the relation in N_k we obtain:

$$\partial_t[dh \wedge \omega_k]_{N_k} = [d\omega_k - df \wedge \omega_k \partial_r]_{N_k}. \tag{5.3}$$

As $d\omega_k$ and $df \wedge \omega_k$ are elements of the kernel of $dh\wedge$, we conclude from Lemma 5.1.5 that the equation (5.3) is equal to

$$\partial_t[\omega_k]_{M_k} = [\frac{d\omega_k}{dh} - \frac{df \wedge \omega_k}{dh}\partial_r]_{M_k},$$

where $[\frac{\omega}{dh}]_{M_k}$ denotes the inverse of $[\omega]_{N_k}$ for $\omega \in \ker^{n+k} dh\wedge$.
Because h is smooth outside D we can lift the vector field $\partial_t \in \Theta_T$ to a vector field $X \in \Theta_V(*D)$, i.e. $X(h) = 1$.
Now for any $\omega = dh \wedge \eta \in \ker^{n+k} dh\wedge$ we have in Ω_V^{n-1+k}

$$\iota_X(\omega) = \iota_X(dh \wedge \eta) = \eta - dh \wedge \iota_X(\eta).$$

Hence in $\Omega^{n-1+k}_{V/T}$ we have $[\eta]_{\Omega_{V/T}} = [\iota_X(\omega)]_{\Omega_{V/T}}$. We recall that any element ω of $\ker^{n+k} dh\wedge$ can be written as $\omega = dh \wedge \eta$ and $[\frac{\omega}{dh}]_{\Omega_{V/T}} = [\eta]_{\Omega_{V/T}}$.
As $d\omega_k \in \ker(dh\wedge)$, we have $d\omega_k = dh \wedge \eta$ for some η, so we conclude

$$[\iota_X(d\omega_k)]_{\Omega_{V/T}} = [\frac{d\omega_k}{dh}]_{\Omega_{V/T}}$$

and in the same way we conclude

$$[\frac{df \wedge \omega_k}{dh}]_{\Omega_{V/T}} = [\iota_X(df)\omega_k - df \wedge \iota_X(\omega_k)]_{\Omega_{V/T}}.$$

So we put this together and use the definition $\mathrm{Lie}_X(\omega) = \iota_X d\omega + d\iota_X\omega$ and the relation $d\omega = df \wedge \omega \partial_r$ in M_k :

$$\begin{aligned}
\partial_t[\omega_k]_{M_k} &= [\frac{d\omega_k}{dh} - \frac{df \wedge \omega_k}{dh}\partial_r]_{M_k} \\
&= [\iota_X(d\omega_k) - \iota_X(df)\omega_k\partial_r + df \wedge \iota_X(\omega_k)\partial_r]_{M_k} \\
&= [\iota_X(d\omega_k) - \mathrm{Lie}_X(f)\omega_k\partial_r + d\iota_X(\omega_k)\partial_r]_{M_k} \\
&= [\mathrm{Lie}_X(\omega_k) - \mathrm{Lie}_X(f)\omega_k\partial_r]_{M_k}
\end{aligned}$$

As h is homogeneous, we can take $X = \frac{E}{n_h \cdot h}$ with $n_h = \deg(h)$ and $E = \sum x_i \partial_i$. We conclude that the $\mathcal{D}_{R\times T}$−actions on M_k is given by :

$$\partial_r.[\omega] = [\omega \otimes \partial_r] \tag{5.4}$$

$$\partial_t.[\omega] = \frac{1}{n_h t}[Lie_E(\omega) - Lie_E(f)\omega \otimes \partial_r]. \tag{5.5}$$

We summarize the discussion:

Proposition 5.1.8. The localized Gauss-Manin system $\Phi_+ \mathcal{O}_V(*D)$ is represented by the complex $(\Phi_* \Omega_{V/T}^{n-1+\bullet}(*D)[\partial_r], d - df \wedge \lrcorner\partial_r)$ and the $\mathcal{D}_{R\times T}$−action on $\mathrm{H}^k(\Phi_* \Omega_{V/T}^{n-1+\bullet}(*D)[\partial_r], d - df \wedge \lrcorner\partial_r)$ is given by

$$\begin{aligned} \partial_r.[\omega] &= [\omega \otimes \partial_r] \\ \partial_t.[\omega] &= \frac{1}{n_h t}[Lie_E(\omega) - Lie_E(f)\omega \otimes \partial_r]. \end{aligned}$$

5.2 The partial Fourier Laplace transform of the Gauss-Manin system

Let us begin by remembering the Fourier transform of a module over the Weyl algebra in one variable $D_1 = \mathbb{C}[r]\langle\partial_r\rangle$, as it is introduced in [Sab06] chapter V. For the general setting we refer to [Sab97] §1 and §2.
The map

$$\begin{aligned} \mathbb{C}[r]\langle\partial_r\rangle &\to \mathbb{C}[\tau]\langle\partial_\tau\rangle \\ r &\mapsto -\partial_\tau \\ \partial_r &\mapsto \tau \end{aligned}$$

is an isomorphism of algebras, because the defining relation $[\partial_r, r] = 1$ in $\mathbb{C}[r]\langle\partial_r\rangle$ can also be written as $[(-r), \partial_r] = 1$. By this isomorphism a (left) $\mathbb{C}[r]\langle\partial_r\rangle$−module M becomes a (left) $\mathbb{C}[\tau]\langle\partial_\tau\rangle$−module, which we denote by $\widehat{M}$.
It is known that if a holonomic $\mathbb{C}[r]\langle\partial_r\rangle$−module M has a regular singularity at infinity, then its Fourier transform $\widehat{M}$ has singularities only at $\tau = 0$ and $\tau = \infty$, where the first is regular and the second has Poincaré rank less than or equal to 1.

We now specialize the situation of the previous section to the following one:

Let h be an equation of the discriminant of a reductive prehomogeneous vector space (G, V) and $f \in V^* \backslash D^*$ be a generic linear form.

In particular h is homogeneous by lemma 1.1.8. We set $n_h := \deg(h)$. As in the previous section we consider $h\colon V \to T = \mathrm{Spec}(\mathbb{C}[t])$ and $f\colon V \to R = \mathrm{Spec}(\mathbb{C}[r])$ as morphisms of affine varieties and set

$$M := \frac{\Phi_*\Omega^{n-1}_{V/T}(*D)[\partial_r]}{(d - df \wedge \lrcorner\partial_r)\Phi_*\Omega^{\bullet+n-2}_{V/T}(*D)[\partial_r]}.$$

M has the structure of a holonomic $\mathcal{D}_{R\times T}-$module since it is the top cohomology group $\mathrm{H}^0(\Phi_+\mathcal{O}_V(*D))$, as shown in proposition 5.1.8. As usual, we will also consider it as a module over $\mathbb{C}[r,t]\langle\partial_r,\partial_t\rangle$. Notice that since we work with forms meromorphic along D, the module M is localized along $t=0$.
We will consider an variant of the above construction, called *partial Fourier-Laplace transformation.* It produces a module $\widehat{M}$ which is equal to M as $\mathbb{C}[t]\langle\partial_t\rangle$-module and where we put $\tau\cdot := \partial_r\cdot$ and $\partial_\tau\cdot := -r\cdot$. Then $\widehat{M}$ becomes a (holonomic) $\mathbb{C}[\tau,t]\langle\partial_\tau,\partial_t\rangle$-module. We consider the localization $\widehat{M}[\tau^{-1}]$, which is called localized Fourier-Laplace transformation of M. It is convenient to write $\theta = \tau^{-1}$. This yields the following definition.

Definition 5.2.1. We call the $\mathbb{C}[t,t^{-1},\theta,\theta^{-1}]-$module

$$G(*D) := \frac{h_*\Omega^{n-1}_{V/T}(*D)[\theta,\theta^{-1}]}{(\theta d - df\wedge)h_*\Omega^{n-2}_{V/T}(*D)[\theta,\theta^{-1}]}$$

the localized Fourier-Laplace transformed Gauss-Manin system, FL-Gauss-Manin system for short, of (f,h) and the $\mathbb{C}[t,\theta]-$module

$$G_0(log\ D) := \frac{h_*\Omega^{n-1}_{V/T}(log\ D)[\theta]}{(\theta d - df\wedge)h_*\Omega^{n-2}_{V/T}(log\ D)[\theta]}$$

the FL-Brieskorn lattice of (f,h).

Remark 5.2.2. As we will see later, $G(*D)$ (resp. $G_0(log\ D)$) are free modules over $\mathbb{C}[t,t^{-1},\theta,\theta^{-1}]$ (resp. over $\mathbb{C}[t,\theta]$), hence they correspond to algebraic bundles over $\mathbb{C}^2$. Because of $\theta = \tau^{-1}$, we have $\theta^2\partial_\theta = -\partial_\tau = r$. We recall that r acts by multiplication with f. Hence $G(*D)$ carries the following $\mathbb{C}[t,\theta]\langle\partial_\theta,\partial_t\rangle-$structure by the equations (5.4) and (5.5):

$$\begin{aligned} \partial_\theta.\omega &= \theta^{-2}f\omega \\ \partial_t.\omega &= \frac{1}{n_h t}(Lie_E(\omega) - \theta^{-1}f\omega). \end{aligned}$$

Here E denotes the Euler field $E = \sum x_i \partial_i$, so $Lie_E(f) = f$.
Notice that this structure induces a $\mathbb{C}[t,\theta]\langle\theta^2\partial_\theta, t\theta\partial_t\rangle$−structure on the FL-Brieskorn lattice $G_0(log\ D)$.

We will now bring us into the situation of [GMS09] §4. This begins by proving the following lemma:

Lemma 5.2.3. (Division lemma - cf.[GMS09] Lemma 4.3) Let D be the discriminant in a reductive prehomogeneous vector space. We have in $G_0(log\ D)$ for all $g \in \mathcal{O}_V$ and all $\xi \in \mathrm{Der}(-log\ h)_0$:

$$g\xi(f)\alpha = \theta\xi(g)\alpha.$$

Proof. Before we will prove the statement, we remark:
If $\xi = \sum_{j=1}^n g_j(x)\partial_j \neq 0$ is a vector field of weight zero, then we can write $\xi = \sum_{i,j=1}^{n} x_i A_{ij} \partial_j$ for some matrix $A = (A_{ij}) \in \mathrm{Mat}_{n\times n}(\mathbb{C})$.
We define $\mathrm{trace}(\xi) := \mathrm{trace}(A)$ and calculate:

$$\begin{aligned}
Lie_\xi(vol) &= d\iota_\xi(vol) = d(\sum_{l=1}^{n}(-1)^{l+1}\iota_\xi(dx_l)\widehat{dx_l} \\
&= d(\sum_{i=1}^{n}(-1)^{l+1}\sum_{l=1}^{n} x_i A_{il}\widehat{dx_l} \\
&= \sum_{i=1}^{n}\sum_{l=1}^{n}(-1)^{l+1}A_{il}dx_i \wedge \widehat{dx_l} = \sum_{l=1}^{n} A_{ll} vol \\
&= \mathrm{trace}(A) vol.
\end{aligned}$$

Suppose ξ annihilates h, then we have $\mathrm{Lie}_\xi(\frac{vol}{h}) = \frac{\mathrm{Lie}_\xi(\mathrm{vol})}{h} = \mathrm{trace}(\xi)\frac{vol}{h}$.
Moreover we have by lemma 3.2.5 for all $g \in \mathcal{O}_V$ and $\xi \in \mathrm{Der}(-log\ h)_0$ in $\Omega^{n-1}_{V/T}(log\ D)$:

$$\xi(g)\alpha = \iota_\xi(dg)\alpha = dg \wedge \iota_\xi(\alpha).$$

We use this formula in the second and last step of the following calculation in $G_0(log\ D)$:

$$\begin{aligned}
g\xi(f)\alpha &= gi_\xi(df) \wedge \alpha \\
&= g(df \wedge i_\xi(\alpha)) \\
&= \theta d(gi_\xi(\alpha)) \\
&= \theta(dg \wedge i_\xi(\alpha) + gdi_\xi(\alpha)) \\
&= \theta(\xi(g) + g \cdot \mathrm{trace}(\xi))\alpha.
\end{aligned}$$

To finish the proof we recall from chapter 4 that $\mathfrak{a} = \mathrm{Der}(-log\ h)_0$, where the Lie algebra $\mathfrak{a}$ was defined as the kernel of the differential $d\chi_h$ of the character χ_h of the semi-invariant h. By [GS10] theorem 2.4 we have $\mathfrak{a} \subset \mathrm{Lie}(\mathrm{SL}(V))$, so we conclude $\mathrm{trace}(\xi) = 0$ for all $\xi \in \mathrm{Der}(-log\ h)_0$. Together with the calculation from above this finishes the proof. □

Proposition 5.2.4. Let (G, V) be a regular reductive prehomogeneous vector space. Let h be an equation of the discriminant (reduced or not). For $f \in V^* \backslash D^*$ the module $G(*D)$ (resp. $G_0(log\ D)$) is free over $\mathbb{C}[t, t^{-1}, \theta, \theta^{-1}]$ (resp. $\mathbb{C}[t, \theta]$) and a basis is given by

$$\omega_i := (-f)^i[\alpha],\ i = 0, \ldots, n_h - 1,$$

where $n_h = \deg(h)$.

Proof. The proof is done similar to the one of proposition 8 of [DG07] or lemma 4.2 of [GMS09]. The difference between their and our situation is that we do not have the equality $\deg(h) = \dim(V)$ and that the module $\mathrm{Der}(-log\ h)$ is not necessarily free. Hence we replace the $\mathcal{O}_V$−module $\mathrm{Der}(-log\ h)$ by $\tilde{\mathfrak{a}}$, the $\mathcal{O}_V$−module generated by the image of the Lie algebra $\mathfrak{a}$ under the map $A \mapsto x^T A^T \partial$. Let us begin with some remarks first:

1. Because $G(*D) = \mathbb{C}[t, t^{-1}, \theta, \theta^{-1}] \otimes_{\mathbb{C}[t,\theta]} G_0(log\ D)$, it is sufficient to prove that $G_0(log\ D)$ is generated freely by $(\omega_i)_{i=0,\ldots,n_h-1}$ over $\mathbb{C}[t, \theta]$.

2. $\{\omega_i := (-f)^{i-1}[\alpha] | i = 0, \ldots, n_h - 1\}$ is a basis if and only if $\{f^{i-1}[\alpha] | i = 0, \ldots, n_h - 1\}$ is a basis.

3. As any $\omega \in G_0(log\ D)$ is of the form $\omega = \sum \delta_i \theta^i$ with $\delta_i \in \Omega^{n-1}_{V/T}(log\ D)$, it is enough to proof that every δ_i is of the form $\sum_j b_{ij}(t, \theta)\omega_j$, for some $b_{ij}(t, \theta) \in \mathbb{C}[t, \theta]$, i.e. it is sufficient to consider $\omega \in \frac{G_0(log\ D)}{\theta G_0(log\ D)}$.

Let $\omega \in \Omega^{n-1}_{V/T}(log\ D)$ be a representative of a class

$$[\omega] \in \frac{G_0(log\ D)}{\theta G_0(log\ D)} \cong \frac{\Omega^{n-1}_{V/T}(log\ D)}{df \wedge \Omega^{n-2}_{V/T}(log\ D)}.$$

By lemma 3.2.2 we have $\Omega^{n-1}_{V/T}(log\ D) \cong \mathcal{O}_V[\alpha]$, so we can assume $\omega = g[\alpha]$ for some $g \in \mathcal{O}_V$. Without loss of generality we can assume g to be homogeneous of degree l, otherwise we do the following calculation for each homogeneous component of g. By proposition 3.2.6 we have:

$$\frac{G_0(log\ D)}{\theta G_0(log\ D)} \cong \frac{\mathcal{O}_V}{df(\mathrm{Der}(-log\ h))}[\alpha], \tag{5.6}$$

which is by proposition 4.1.11 freely generated by $1, f, f^2, \ldots, f^{n_h-1}$ as a $\mathbb{C}[t]$−module. Hence we can write g as

$$g = \tilde{g}(h)f^i + \eta(f),$$

where $\tilde{g} \in \mathbb{C}[t]$ is a polynomial of degree $\lfloor \frac{l}{n_h} \rfloor$, $i = l - n_h \deg(\tilde{g})$ and $\eta \in \mathrm{Der}(-log\ h)$ is homogeneous of $\deg(\eta) = l - 1$. Here $\lfloor \cdot \rfloor$ denotes the floor function.
Because $f \in V^* \backslash D^*$ we have $df(\mathrm{Der}(-log\ h)) = df(\mathfrak{a})$ by prop. 4.3.1 and can write

$$\eta(f) = \sum_{j=1}^{n-1} k_j^0 \xi_j(f),$$

where $\xi_1, \ldots, \xi_{n-1}$ is a basis of $\mathfrak{a}$ and k_j^0 are homogeneous polynomials of degree $\deg(g) - 1$. Hence we have

$$[\omega] = [(\tilde{g}f^i + \sum_{j=1}^{n-1} k_j^0 \xi_j(f))\alpha].$$

We apply the division lemma (lemma. 5.2.3) and get in $G_0(log\ D)$:

$$[\omega] = [(\tilde{g}(h)f^i + \theta \sum_{j=1}^{n-1} \xi_j(k_j^0))\alpha].$$

The polynomial $k^1 = \sum_j \xi_j(k_j^0)$ is homogeneous of degree $\deg(g) - 1$ and we can apply the argument again, i.e. we find homogeneous k_j^1 of degree $l - 2$ and a homogeneous $\tilde{g}^1 \in \mathbb{C}[t]$ such that :

$$[k^1\alpha] = [(\tilde{g}^1(h)f^{i-1} + \theta(\sum_{j=1}^{n-1} \xi_j(k_j^1))\alpha].$$

Hence iterating the argument, we see that $(\omega_i = f^{i-1}\alpha)_{i=0,\ldots,n_h-1}$ is a system of generators of $G_0(log\ D)$ as a $\mathbb{C}[t, \theta]$−module.

To prove the freeness, we consider a relation

$$\sum_{j=0}^{n_h-1} k_j(t, \theta) f^j \alpha = (\theta d - df \wedge) \sum_{i=l}^{L} \eta_i \theta^i$$

where $\eta_i \in \Omega^{n-2}_{V/T}(log\ D)$ and $0 \leq l \leq L$. Writing the left hand side as a polynomial in θ with coefficients in $\Omega^{n-1}_{V/T}(log\ D)$, the equation becomes

$$\sum_{i=l}^{L+1} \delta_i \theta^i = (\theta d - df\wedge) \sum_{i=l}^{L} \eta_i \theta^i,$$

where $\delta_i = \sum_{j=0}^{n_h-1} b_{ij}(t) f^j \alpha$. We conclude $\delta_l = -df \wedge \eta_l \in df \wedge \Omega^{n-2}_{V/T}(log\ D)$, i.e. $\delta_l = [0] \in \frac{\Omega^{n-1}_{V/T}}{df\wedge\Omega^{n-2}_{V/T}}$, but by proposition 4.1.11 $([f^j\alpha])_{j=0,\ldots,n-1}$ is a basis of $\frac{\Omega^{n-1}_{V/T}(log\ D)}{df\wedge\Omega^{n-2}_{V/T}(log\ D)}$, so $b_{l,j}(t) = 0$ for $j = 0, \ldots, n-1$ and hence $\delta_l = 0$. Iterating the argument, we see that $\delta_i = 0$ for all $i = l, \ldots, L+1$, which shows that all $k_j(t,\theta) = 0$ and that the relation is trivial. □

As the modules $G(*D)$ and $G_0(log\ D)$ are free, they correspond to vector bundles on $\mathbb{C}_\theta \times T$ with a connection, which we denote by $(G(*D), \nabla)$ and $(G_0(log\ D), \nabla)$. Moreover by the formulas 5.4 and 5.5 the relative connections are

$$\begin{aligned} \nabla_{\partial_\theta}[\omega] &= [\theta^{-2} f\omega] \\ \nabla_{\partial_t}[\omega] &= \frac{1}{n_h t}[L_E(\omega) - \theta^{-1} f\omega]. \end{aligned}$$

Of course the poles of the connection

$$\nabla\colon G_0(log\ D) \to G_0(log\ D) \otimes \theta^{-1}\Omega^1_{\mathbb{C}\times T}(log(\{0\}\times T) \cup (\mathbb{C}\times\{0\})).$$

are complicated, but as we are now in the situation[2] of [GMS09] we have by the reductivity of the group G

Proposition 5.2.5. (cf. [GMS09] Proposition 4.5)

1. There is a basis $\omega^{(1)} \subset G_0(log\ D)$, represented by differential forms $\omega^{(1)}_i = [g_i\alpha]$ with g_i homogeneous of degree i, such that

$$\nabla\omega^{(1)} = \omega^{(1)}[(A_0\frac{1}{\theta} - A_\infty)\frac{d\theta}{\theta} + (-A_0\frac{1}{\theta} - A_\infty + D)\frac{dt}{n_h t}],$$

[2]Except that we do not have $\deg(h) = \dim(V)$. Hence in our situation we should keep in mind that the rank of $G_0(log\ D)$ is equal to $n_h = \deg(h)$ and not to $\dim(V)$

where the $n_h \times n_h-$matrices are

$$A_0 = \begin{pmatrix} 0 & 0 & 0 & \cdots & 0 & c\cdot t \\ 1 & 0 & 0 & \cdots & 0 & 0 \\ 0 & 1 & 0 & \cdots & 0 & 0 \\ 0 & 0 & 1 & \cdots & 0 & 0 \\ \vdots & \vdots & \vdots & \cdots & \vdots & \vdots \\ 0 & 0 & 0 & \cdots & 1 & 0 \end{pmatrix}$$

and $A_\infty = \operatorname{diag}(\nu_1, \dots, \nu_{n_h})$, where the constant $c \in \mathbb{C}$ is defined by the equation $(-f)^{n_h} = \sum_{i=1}^{n_h} g_i \xi_i(f) + c \cdot h$ with $g_i \in \mathcal{O}_V$ and $\xi \in \tilde{\mathfrak{a}}$.

2. The connection is flat outside $\theta = 0$ and $t = 0$.

By the concrete formulas for the relative connections, we see that

$$\theta^2 \partial_\theta = \theta \mathrm{Lie}_E(\omega) - nt\theta\partial_t,$$

where we write $\partial_\theta = \nabla_{\partial_\theta}$ and $\partial_t = \nabla_{\partial_t}$. So if we knew the connection of one of the restriction $G_t := (G_t(*D), \nabla_{|t})$ or $G_\theta := (G_\theta(*D), \nabla_{|\theta})$, we knew also the other one. In particular the restriction $G_1(*D)$ to $\theta = 1$ has a regular singularity at $t = 0$ and obviously the spectrum of the matrix $(-A_0(0) - A_\infty + D)$ is equal to the spectrum of the residue endomorphism.
Moreover, as discussed in section 2.2, if we have the basis $\omega^{(1)}$ we can always produce a V^+-solution $\omega^{(3)}$ by applying the algorithms 1 and 2.

We summarize this chapter:

Theorem 5.2.6. Let (G, V) be a regular reductive prehomogeneous vector space with discriminant D. Let h be an equation of degree $n_h = \deg(h)$ for D (reduced or not) and $f \in V^* \backslash D^*$ be any generic linear form. Then there is a $\mathbb{C}[\theta, t]-$basis $\omega^{(3)} = (\omega_1^{(3)}, \ldots, \omega_{n_h}^{(3)})$ of the FL-Brieskorn lattice $G_0(log\ D)$, such that

$$\nabla \omega^{(3)} = \omega^{(3)} \left[(A_0 \theta^{-1} + A_\infty) \frac{d\theta}{\theta} + (-A_0 \theta^{-1} - A_\infty + D) \frac{dt}{n_h t} \right]$$

where

$$A_0 = \begin{pmatrix} 0 & 0 & \ldots & 0 & c \cdot t \\ 1 & 0 & \ldots & 0 & 0 \\ \vdots & \vdots & \ldots & \vdots & \vdots \\ 0 & 0 & \ldots & 0 & 0 \\ 0 & 0 & \ldots & 1 & 0 \end{pmatrix}$$

$A = \mathrm{diag}(\nu_1, \ldots, \nu_{n_h})$ and $D = \mathrm{diag}(0, 1, \ldots, n_h - 1)$. Moreover $(\nu_1, \ldots, \nu_{n_h})$ is the spectrum at infinity of the restriction $G_0(log\ D)_{|t}$.

Chapter 6

The Bernstein-Sato polynomial

We already met the Bernstein-Sato polynomial in the section about prehomogeneous vector spaces in the way it was introduced by Sato. But the defining functional equation holds in a more general situation (see [Ber72] theorem 1' for algebraic functions and [Bjö79] Ch. 3 cor. 3.6 for analytic functions):

Theorem 6.0.1. For any function $h \in \mathcal{O}_V$ there exists a non zero polynomial $b[s] \in \mathbb{C}[s]$ and a differential operator $P(x_i, \partial_i, s) \in \mathcal{D}_V[s]$, such that the following functional equation holds

$$P(x_i, \partial_i, s)h^{s+1} = b(s)h^s.$$

Definition 6.0.2. For $h \in \mathcal{O}_V$ we denote by $b_h(s)$ the unitary generator of the ideal defined by the functional equation of theorem 6.0.1 and call it the *Bernstein-Sato polynomial of h.*

By [Kas76] §5 the roots of the Bernstein-Sato polynomial are negativ rational numbers and by [Mal74] there is a close connection between Bernstein-Sato polynomial of h and the local monodromy of the zero fibre $h^{-1}(0)$. Precisely if $h\colon \mathbb{C}^n \to \mathbb{C}$ has an isolated singularity at the origin, then by [Mal74] the eigenvalues of local monodromy are $\exp(2\pi i \lambda)$, where the lambdas are the roots of b_h.

In general not much is known about the operator P in the functional equation, but when h is an equation for the discriminant D of a regular reductive prehomogeneous vector space (G, ρ, V), then this operator is of a special type. We recall from section 1:
By theorem 1.3.2 the dual triple (G, ρ^*, V^*) is a prehomogeneous vector

space and (in unitary coordinates) $h^*(y) := \overline{h(\overline{y})}$ is an equation for the dual discriminant D^*. Moreover there is a polynomial $b_h(s)$, which we called the b–function of h, satisfying the functional equation :

$$h^*(\partial)h^{s+1} = b_h(s)h^s.$$

In particular the b–function is minimal with respect to degree and we have $\deg(b_h(s)) = \deg(h)$. Moreover any semi-invariant h of the triple (G, ρ, V) fulfils such a functional equation.

In this chapter we will relate roots of the Bernstein-Sato polynomial of h to the spectrum of the FL-Brieskorn lattice (cf. theorem 5.2.6).

Corollary. (Corollary 6.2.10) Let h be an equation of the discriminant in a regular reductive prehomogeneous vector space. We denote by $\nu_1, \ldots, \nu_{\deg(h)}$ the spectrum of the FL-Brieskorn lattice and by $\alpha_1, \ldots, \alpha_{\deg(h)}$ the roots of the Bernstein-Sato polynomial of h. Then we have

$$\alpha_i = \frac{i - 1 - \nu_i}{\deg(h)} - 1.$$

Therefore we will recall in section 6.1 the definition of the Bernstein-Sato polynomial by the V–filtration and relate them to the classical definition. Then we will introduce in analogy to [Mal83] different cyclic $\mathcal{D}$–modules and relate the Bernstein-Sato polynomials of their generators in section 6.2.

6.1 The Bernstein-Sato polynomial of the V–filtration

In this section we will give a description of the Bernstein-Sato polynomial $b_h(s)$ by the V–filtration. For the convienice of the reader we give a brief introduction to the theory of the V–filtration for $\mathcal{D}$–modules and refer to [MM04] §4 or [Kas83] for more details.

Let V be an affine space of dimension n and $T = \mathrm{Spec}(\mathbb{C}[t])$. We consider V as a hypersurface in

$$X = T \times V \cong \mathbb{C}^{n+1}.$$

The *V–filtration of $\mathcal{D}_X$ (with respect to the hypersurface V)* is by definition the filtration

$$V_k\mathcal{D}_X := \{P \in \mathcal{D}_X \mid \forall l \in \mathbb{Z} : P\mathcal{J}^{l+k} \subset \mathcal{J}^l\},$$

where $\mathcal{J} = (t)$ and $\mathcal{J}^k = \mathcal{O}_X$ for $k \leq 0$. It is an increasing and exhaustive filtration of $\mathcal{D}_X$.

In local coordinates we can write a differential operator $P \in \mathcal{D}_X$ uniquely as

$$P = \sum_{finite} g_{\alpha,i,j} \partial_x^\alpha \partial_t^i t^j$$

with $g_{\alpha,i,j} \in \mathcal{O}_{T\times V}$ and $\partial_x^\alpha = \partial_1^{\alpha_1} \dots \partial_n^{\alpha_n}$ for $\alpha = (\alpha_1, \dots, \alpha_n)$ and by definition of the $V-$filtration we have

$$-\partial_t^l t^k \in V_{l-k}\mathcal{D}_X, \text{ for all } k, l \in \mathbb{N}^2.$$

Hence we have for $k \in \mathbb{N}$

$$\begin{aligned} V_{-k}\mathcal{D}_X &= \{P = \sum_{finite} g_{\alpha,j} \partial_x^\alpha (-\partial_t t)^j t^k,\ g_{\alpha,j} \in \mathcal{O}_X\} \\ &= \mathcal{D}_V[t]\langle -\partial_t t\rangle t^k \\ V_k\mathcal{D}_X &= \{P = \sum_{l=0}^{k} \sum_{finite} g_{\alpha,j,l} \partial_x^\alpha \partial_t^l (-\partial_t t)^j,\ g_{\alpha,j,l} \in \mathcal{O}_X\}. \end{aligned}$$

Definition 6.1.1. Let $\mathcal{M}$ be a coherent $\mathcal{D}_X-$module. We call a increasing exhaustive filtration $U_k\mathcal{M}$ by coherent $V_0\mathcal{D}_X-$modules *good*, if the filtration is compatible the filtration $V_k\mathcal{D}_X$, i.e.

$$V_k\mathcal{D}_X U_l\mathcal{M} \subset U_{l+k}\mathcal{M}$$

and there exists a $k_0 \in \mathbb{N}$ such that for all $k \in \mathbb{N}$:

$$V_k\mathcal{D}_X U_{k_0}\mathcal{M} = U_{k+k_0}\mathcal{M} \text{ and } V_{-k}\mathcal{D}_X U_{-k_0}\mathcal{M} = U_{-k-k_0}\mathcal{M},$$

in other words $V_k\mathcal{D}_X U_l\mathcal{M} \subset U_{l+k}\mathcal{M}$ is an equality almost all k.

Proposition 6.1.2. (cf. [MM04] 4.2-2) For a coherent $\mathcal{D}_X-$module the following statements are equivalent

1. For all good $V-$filtrations $U_\bullet\mathcal{M}$ there exists a non zero polynomial $b(s) \in \mathbb{C}[s]$, such that $b(-\partial_t t + k)U_k\mathcal{M} \subset U_{k-1}\mathcal{M}$ holds for all $k \in \mathbb{Z}$.

2. There exists a good $V-$filtrations $U_\bullet\mathcal{M}$ and a non zero polynomial $b(s) \in \mathbb{C}[s]$, such that $b(-\partial_t t + k)U_k\mathcal{M} \subset U_{k-1}\mathcal{M}$ holds for all $k \in \mathbb{Z}$.

3. For all finite systems $\{m_1, \dots, m_n\}$ of generators of $\mathcal{M}$ there exists a non zero polynomial $b(s) \in \mathbb{C}[s]$ such that $b(-\partial_t t)m_i \in \sum_{j=1}^n V_{-1}\mathcal{D}_X m_j$.

A $\mathcal{D}_X$-module $\mathcal{M}$ is called *specializable along the hypersurface* V, if one of the properties of the previous proposition is satisfied. If $\mathcal{M}$ is a holonomic $\mathcal{D}_X$-module, then by proposition 4.4-2 [MM04] $\mathcal{M}$ is specializable along (any hypersurface) V.
As all the module, we consider, are holonomic, we define :

Definition 6.1.3. Let $\mathcal{M}$ be a (holonomic) coherent $\mathcal{D}_X$-module and $U_\bullet\mathcal{M}$ be a good V-filtration. We call the unitary generator of the ideal generated by the polynomials $b(s) \in \mathbb{C}[s]$ satisfying

$$b(-\partial_t t + k)U_k\mathcal{M} \subset U_{k-1}\mathcal{M}, \ \forall k \in \mathbb{Z}$$

the *Bernstein-Sato polynomial of the good V-filtration $U_\bullet\mathcal{M}$.*
We denote this Bernstein-Sato polynomial by $b_U(s)$.
For a section $m \in \mathcal{M}$ we define the *Bernstein-Sato polynomial of the section m* as the unitary generator of the ideal defined by

$$b(-\partial_t t)m \in V_{-1}\mathcal{D}_X m.$$

We denote this Bernstein-Sato polynomial by $b_m(s)$.
We will now connect these a priori different definitions of the Bernstein-Sato polynomial for the datum of a regular function $h\colon V \to T = \mathbb{C}_t$. Therefore we consider s as an extra parameter and the $\mathcal{O}_V[h^{-1}, s]$-module $\mathcal{M}[h^{-1}] = \mathcal{O}_V[h^{-1}, s]h^s$ generated by the formal symbol h^s. This module becomes a $\mathcal{D}_X$-module by

$$\begin{aligned}
\partial_i(gh^s) &:= \partial_i(g)h^s + sg\frac{\partial_i(h)}{h}h^s \\
t(g(s)h^s) &:= g(s+1)h^{s+1} \\
\partial_t(g(s)h^s) &:= -sg(s-1)h^{-1}h^s.
\end{aligned}$$

We remark that for $P \in \mathcal{D}_V$ and $\phi(t) \in \mathbb{C}[t]$ we have $[\phi, P] = 0$ and $\phi s = s\phi + t\phi'$. Indeed the first statement follows from a direct calculation and by definition we have $(s+1)t.(g(s)h^s) = t.(s(g(s)h^s))$, so the second statement follows by induction.

Lemma 6.1.4. The map $t\colon \mathcal{M}[h^{-1}] \to \mathcal{M}[h^{-1}]$ is bijective.

Proof. The injectivity and the surjectivity are following from t-action. For the surjectivity we remark that we have by definition

$$g(s)h^s = t.(\frac{g(s-1)}{h}h^s).$$

□

Using the formula for the inverse t^{-1} and the commutation rule for t and s one proves the following lemma, which verifies the ∂_t−action :

Lemma 6.1.5. $\nabla_{\partial_t} := -t^{-1}(s+1)$ defines a connection on $\mathcal{M}[h^{-1}]$. In particular we have $-t\partial_t = (s+1)$ and $-\partial_t t$ acts as multiplication with s.

We denote by $\mathcal{M}$ the submodule $\mathcal{M} := \mathcal{D}_{T\times V}h^s$ in $\mathcal{M}[h^{-1}]$ and filter this modul by

$$U_k\mathcal{M} := V_k\mathcal{D}_{T\times V}h^s.$$

Remark 6.1.6. The filtration $U_\bullet\mathcal{M}$ is a good V−filtration and the localization of $\mathcal{M}$ with respect to h is equal to $\mathcal{M}[h^{-1}]$.

Lemma 6.1.7. For a polynomial $b(s) \in \mathbb{C}[s]$ holds

$$b(s+1)h^s = P(s)h^{s+1},$$

for some $P(s) \in \mathcal{D}_V[s]$, if and only if

$$b(-t\partial_t)h^s \in V_{-1}\mathcal{D}_{T\times V}h^s.$$

Proof. If $b(-t\partial_t)h^s \in V_{-1}\mathcal{D}_X h^s$, then $b(-t\partial_t) = P(-t\partial_t)th^s$ and by $(s+1) = -t\partial_t$ we get $b(s+1)h^s = \tilde{P}(s)h^{s+1}$, with $\tilde{P}(s) = P(s+1)$.
Conversely using $(s+1) = -t\partial_t$, we write $b(s+1)h^s = P(s)h^{s+1}$ as $b(-t\partial_t)h^s = P(s)t.h^s \in V_{-1}D_{T\times V}h^s$. □

Corollary 6.1.8. The Bernstein-Sato polynomial b_h of h and the Bernstein-Sato polynomial b_{h^s} of the section $h^s \in \mathcal{M}$ are equal and as h^s is a generator of $\mathcal{M}$ both are equal to the Bernstein-Sato polynomial $b_U(s)$ of the good V−filtration $U_\bullet\mathcal{M}$ of $\mathcal{M}$.

6.2 A comparision of the Bernstein-Sato polynomial with the spectral polynomial

It was shown in [Sev11] §3 that if $D = \mathcal{V}(h)$ is a reductive linear free divisor, then the Bernstein-Sato $b_h(s)$ polynomial of h and the spectral polynomial $b_{G_1(log\ D)}(s)$ fulfil

$$b_{G_1(log\ D)}(s+1) = b_h(s).$$

We will use the ideas of [Sev11] and prove the same result in our more general situation :
Let (G, V) be a regular reductive prehomogeneous vector space. Let h be an equation (reduced or not) of the complement of the open orbit of degree

$$\deg(h) = n_h.$$

Furthermore we consider h as a morphism

$$h \colon V \to T = \mathrm{Spec}(\mathbb{C}[t])$$

and choose a generic $f \in V^* \backslash D^*$. As we have seen the FL-Gauss-Manin system

$$G(*D) = \frac{h_* \Omega^{n-1}_{V/T}(*D)[\theta, \theta^{-1}]}{(\theta d - df \wedge) h_* \Omega^{n-2}_{V/T}(*D)[\theta, \theta^{-1}]}$$

corresponds to a rational vector bundle on $\mathbb{C}_\theta \times T$ with a meromorphic connection, which we have denoted by $(G(*D), \nabla)$.
Moreover there is a $\mathbb{C}[\theta, t]-$basis $\omega = (\omega_1, \ldots, \omega_{n_h})$ of the FL-Brieskorn lattice $G_0(log\ D)$, such that the connection ∇ on $(G(*D), \nabla)$ is given by

$$\nabla(\omega) = \omega[(A_0 + \theta A_\infty)\frac{d\theta}{\theta^2} + (-A_0 + \theta A'_\infty)\frac{dt}{n_h \theta t}],$$

where

$$\begin{aligned}
A_\infty &= \mathrm{diag}(\nu_1, \ldots, \nu_{n_h}), \\
A'_\infty &= \mathrm{diag}(0, \ldots, n_h - 1) - A_\infty, \\
A_0 &= \begin{pmatrix} 0 & 0 & \ldots & 0 & c \cdot t \\ -1 & 0 & \ldots & 0 & 0 \\ \ldots & \ldots & \ldots & \ldots & \ldots \\ 0 & 0 & \ldots & 0 & 0 \\ 0 & 0 & \ldots & -1 & 0 \end{pmatrix} \quad \text{and} \\
\omega_1 &= \iota_E(\frac{vol}{h}) = n_h \frac{vol}{dh}.
\end{aligned}$$

We remark that because the bundle is invariant under $\nabla_{-n_h \theta t \partial_t + \theta^2 \partial_\theta}$ and has the above properties on the poles it is by [Her03] lemma 7.19 determined by its restrictions to $\theta = 1$.
Let $j \colon \{1\} \times T \to \mathbb{C}_\theta \times T$ be the canonical inclusion. We set

$$(G_1(*D), \nabla) := j^*(G(*D), \nabla),$$

which can be seen as a coherent and holonomic left $\mathbb{C}[t]\langle \partial_t \rangle-$module.

Lemma 6.2.1. We have an isomorphism of left $\mathbb{C}[t]\langle\partial_t\rangle-$modules:

$$\phi\colon \frac{\mathbb{C}[t]\langle\partial_t\rangle}{(b_{G_1(log\ D)}(t\partial_t)-\frac{c}{n_h^{n_h}}\cdot t)} \to (G_1(*D),\nabla)$$
$$[1] \mapsto \omega_1$$

where $b_{G_1(log(D)}(s) := \prod_{i=1}^{n_h}(s-\frac{i-1-\nu_i}{n_h})$ is the spectral polynomial.

Proof. Let ω be the basis of $G_0(log\ D)$ from above. We set $\omega_i' := \omega_{i|\theta=1}$.[1] In this basis the connection on $(G_1(*D),\nabla)$ is given by $\nabla(\omega')=\omega'(-A_0+A_\infty')\frac{dt}{n_h t})$. We write $t\partial_t = \nabla_{t\partial_t}$ and conclude

$$\frac{\omega_{i+1}'}{n_h} = (t\partial_t - \frac{i-1-\nu_i}{n_h})\omega_i',\ \text{for}\ 1\leq i<n_h \qquad (6.1)$$
$$\frac{ct\omega_1'}{n_h} = (t\partial_t - \frac{n_h-1-\nu_{n_h}}{n_h})\omega_{n_h}'. \qquad (6.2)$$

Now we insert the equations (6.1) into the equation (6.2) and see that ω_1' satisfies the equation

$$\frac{ct\omega_1'}{n_h^{n_h}} = b_{G_1(log\ D)}(t\partial_t)\omega_1'.$$

In particular this relation is minimal and $\phi([1])=\omega_1'$ defines the isomorphism, as $\frac{\mathbb{C}[t]\langle\partial_t\rangle}{(b_{G_1(log\ D)}(t\partial_t)-\frac{c}{n_h^{n_h}}\cdot t)}$ is generated as left $\mathbb{C}[t]\langle\partial_t\rangle-$module by the class [1]. □

On the other hand the restriction

$$G_1(*D) = \frac{h_*\Omega_{V/T}^{n-1}(*D)}{(d-df\wedge)h_*\Omega_{V/T}^{n-2}(*D)}.$$

is precisely the direct image under h of the module $\mathcal{O}_V(*D)$ twisted by e^{-f}, that is : For $f\in\mathcal{O}_V$, we denote by $\mathcal{O}_V e^{-f}$ the $\mathcal{O}_V-$module formally generated by e^{-f}.
We define a $\mathcal{D}_V-$structure on $\mathcal{O}_V(*D)e^{-f} := \mathcal{O}_V(*D)\otimes_{\mathcal{O}_V}\mathcal{O}_V e^{-f}$ by

$$\partial_i.(ge^{-f}) := \partial_i(g)e^{-f} - g\partial_i(f)e^{-f},$$

i.e. the $\mathcal{D}_V-$structure defining differential of $\mathcal{O}_V(*D)e^{-f}$ is $\delta = d-df\wedge$.
The direct image $h_+\mathcal{O}_V(*D)e^{-f}$ is represented by the complex

$$(h_*\Omega_V^{n+\bullet}(*D)[\partial_t], d-df\wedge-(dh\wedge\lrcorner\partial_t)),$$

[1] Apart from this proof we will identify the bases ω and ω'.

which is by same arguments as in the proof of proposition 5.1.6 quasi isomorphic to the complex

$$(h_*\Omega^{n-1+\bullet}_{V/T}(*D), d - df\wedge).$$

So we conclude :

Lemma 6.2.2. $G_1(*D)$ is isomorphic to $\mathbb{H}^0(T, h_+\mathcal{O}_V(*D)e^{-f})$ as left $\mathbb{C}[t]\langle\partial_t\rangle-$ module.

We will use this use this module to compare the Bernstein-Sato polynomial with the spectral polynomial later.

The pullback of the FL-Brieskorn lattice $G_1(log\ D) = j^*((G_0(log\ D), \nabla)$ is a $V_0\mathbb{C}[t]\langle\partial_t\rangle-$submodule of $G_1(*D)$ and in particular we have

$$G_1(log\ D) = \mathbb{C}[t]\langle t\partial_t\rangle\omega_1.$$

Lemma 6.2.3. The filtration

$$U_kG_1(*D) := V_k\mathbb{C}[t]\langle\partial_t\rangle \cdot G_1(log\ D)$$

is good and has the Bernstein-Sato polynomial $b_{G_1(log\ D)}(s+1)$.

Proof. $G_1(log\ D)$ is a coherent $\mathcal{O}_T-$module, which is invariant under $t\partial_t$ and therefore it is invariant under $\partial_t t$, so it is a coherent $V_0\mathbb{C}[t]\langle\partial_t\rangle-$module. The filtration $U_\bullet G_1(*D)$ is increasing because the operator ∂_t maps $G_1(log\ D)$ to $\frac{1}{t}G_1(log\ D)$ and by definition of the $V-$filtration on $\mathbb{C}[t]\langle\partial_t\rangle$ we have $tV_k\mathbb{C}[t]\langle\partial_t\rangle = V_{k-1}\mathbb{C}[t]\langle\partial_t\rangle$. Obviously $U_\bullet G_1(*\ D)$ is exhaustive and it is good, because the $t-$action is bijective.

The $V_0\mathbb{C}[t]\langle\partial_t\rangle-$module $G_1(log\ D)$ is generated by ω_1 and by lemma 6.2.1 we have :

$$\frac{c}{n_h^{n_h}}t\omega_1 = b_{G_1(log\ D)}(t\partial_t).$$

We recall that $\deg(h) = n_h = \deg(b_{G_1(log\ D)})$, and calculate :

$$\begin{aligned}(-1)^{n_h}\frac{c}{{n_h}^{n_h}}t\omega_1 &= (-1)^{n_h}b_{G_1(log\ D)}(t\partial_t)\omega_1 \\ &= b_{G_1(log\ D)}(-t\partial_t)\omega_1 \\ &= b_{G_1(log\ D)}(-\partial_t t - 1)\omega_1 \in V_{-1}\mathbb{C}[t]\langle\partial_t\rangle\omega_1.\end{aligned}$$

We conclude that $(-1)^{n_h}b_{G_1(log\ D)}(s+1)$ generates the ideal

$$(b(s) \in \mathbb{C}[s] \mid b(-\partial_t t)\omega_1 \in V_{-1}\mathbb{C}[t]\langle\partial_t\rangle\omega_1)\,,$$

otherwise we would get a relation for ω_1 of degree smaller than n_h.

As $b_{G_1(log\ D)}(s+1)$ is the unitary generator of the ideal spanned by the polynomial $(-1)^{n_h}b_{G_1(log\ D)}(s+1)$, it is the Bernstein-Sato polynomial of the filtration $U_\bullet G_1(*D)$. □

We remember from corollary 6.1.8 that the Bernstein-Sato polynomial $b_h(s)$ of h given by the functional equation

$$b_h(s)h^s = Ph^{s+1}$$

is also the Bernstein-Sato polynomial of the generator h^s of

$$\mathcal{M} = \mathcal{D}_{T\times V}h^s.$$

We will now construct other cyclic $\mathcal{D}-$modules, such that the Bernstein-Sato polynomial of their generators will be $b_h(s)$. Therefore we will often use the fact that given an exact sequence of $\mathcal{D}-$modules

$$0 \to \mathcal{M} \to \mathcal{N} \to \mathcal{L} \to 0$$

a Bernstein-Sato polynomial of a section $m \in \mathcal{M}$ is also a Bernstein-Sato polynomial of m seen as a section in $\mathcal{N}$.
The first construction holds more general. Let $h \in \mathcal{O}_V$ be an arbitrary (non-zero) regular function, seen as a morphism

$$h\colon V \to T.$$

We denote by $i_h\colon V \to T \times V$ the graph embedding and by Γ_h the image of i_h. By the Kashiwara equivalence, e.g. [HT07] theorem 1.6.1, the direct image $(i_h)_+\mathcal{O}_V =: \mathcal{N}$ is a $\mathcal{D}_{T\times V}-$module with support on the graph Γ_h, i.e. we have an isomorphism of $\mathcal{D}_{T\times V}-$modules :

$$\mathcal{N} = (i_h)_+\mathcal{O}_V \cong \frac{\mathcal{O}_{T\times V}(*\Gamma_h)}{\mathcal{O}_{T\times V}}.$$

The module on the right hand side is generated over $\mathcal{D}_{T\times V}$ by the class of the function $\frac{1}{t-h}$, which we denote by $\delta(t-h)$ and the $\mathcal{D}_{T\times V}-$structure is given by :

$$\begin{aligned} \partial_{x_i}.\delta(t-h) &= [\frac{\partial_{x_i}(h)}{t-h}]\delta(t-h) \\ \partial_t.\delta(t-h) &= [-\frac{1}{t-h}]\delta(t-h). \end{aligned}$$

We write $\mathcal{N} = \mathcal{D}_{T\times V}\delta(t-h)$ and recall the $\mathcal{D}_{T\times V}-$action on $\mathcal{M} = \mathcal{D}_{T\times V}h^s$ (seen as a submodule of $\mathcal{O}_V[h^{-1}, s]h^s$) :

$$\begin{aligned} x_i.h^s &= x_i \cdot h^s \\ t.(g(s)h^s) &= g(s+1)h \cdot h^s \\ \partial_{x_i}.(gh^s) &= \partial_{x_i}(g)h^s + \frac{sg\partial_{x_i}(h)}{h}h^s \\ \partial_t.(g(s)h^s) &= \frac{-sg(s-1)}{h}h^s. \end{aligned}$$

Proposition 6.2.4. The map

$$\begin{aligned} \phi\colon \mathcal{D}_{T\times V} h^s &\to \mathcal{D}_{T\times V}\delta(t-h) \\ h^s &\mapsto \delta(t-h) \end{aligned}$$

is an injective $\mathcal{D}_{T\times V}-$module homomorphism.

Proof. At first we remark that for $g = g_1(t-h) \in (t-h)$ we have on the one hand

$$[g]\delta(t-h) = [\frac{g}{t-h}] = [g_1] \subset \mathcal{O}_{T\times V}$$

and on the other hand

$$g_1 \cdot (t-h)h^s = g_1(h^{s+1} - h^{s+1}) = 0,$$

so ϕ is an injective and $\mathcal{O}_V-$linear map and because of

$$\begin{aligned} \phi(t.h^s) &= \phi(h\cdot h^s) = h\cdot\phi(h^s) = [\frac{h}{t-h}] \\ &= [\frac{t-h}{t-h} + \frac{h}{t-h}] = [\frac{t}{t-h}] \\ &= t.\phi(h^s) \end{aligned}$$

it is $\mathcal{O}_{T\times V}-$linear. At second we conclude directly from the $\mathcal{D}_{T\times V}-$actions that in $\mathcal{N}$ and in $\mathcal{M}$ holds :

$$\partial_{x_i} = -\partial_{x_i}(h)\partial_t.$$

Therefore it is enough to check one of the operations. But the statement is clear for ∂_{x_i}, so it is clear for ∂_t. □

Remark 6.2.5. As in the previous section the following filtrations are good :

$$\begin{aligned} U_k\mathcal{N} &= V_k\mathcal{D}_{T\times V}\delta(t-h) \\ U_k\mathcal{M} &= V_k\mathcal{D}_{T\times V}h^s. \end{aligned}$$

Corollary 6.2.6. The Bernstein-Sato polynomial $b_h(s)$ of h given by the functional equation $b(s)h^s = Ph^{s+1}$ is the Bernstein-Sato polynomial of the section $\delta(t-h)$.

Remark 6.2.7. Looking at the modules of global sections we notice that the localizations of $\mathcal{N}$ and $\mathcal{M}$ along $t = 0$ are isomorphic to $(i_h)_+\mathcal{O}_V(*D)$. We denote the localisation by $\mathcal{N}[t^{-1}]$. Because $\mathcal{N}$ has no $t-$torsion a Bernstein-Sato polynomial of a section $m \in \mathcal{N}$ is a Bernstein-Sato polynomial of m as a section in $\mathcal{N}[t^{-1}]$.

We will now come back to our special situation, where h is an equation of the discriminant D in a reductive prehomogeneous vector space (G, V). We recall from the discussion of section 1.3 that functional equation for the Bernstein-Sato polynomial is in this case

$$h^*(\partial)h^{s+1} = b_h(s)h^s.$$

Lemma 6.2.8. $b_h(s)$ is a Bernstein-Sato polynomial of the section $\delta(t - h)e^{-f} \in (i_h)_+\mathcal{O}_V(*D)e^{-f}$ with functional equation :

$$(t \cdot h^*(\partial_i + a_i) - b_h(-\partial_t t))\delta(t - h)e^{-f} = 0$$

for $f = \sum_{i=1}^{n_h} a_i x_i$.

Proof. Because h is a semi-invariant, by 6.2.6 $b_h(s)$ is the unitary polynomial of lowest degree satisfying the functional equation:

$$(th^*(\partial_i) - b_h(-\partial_t t))\delta(t - h) = 0 \tag{6.3}$$

in $\mathcal{N} = (i_h)_+\mathcal{O}_V$. This also holds in $\mathcal{M}[t^{-1}] = (i_h)_+\mathcal{O}_V(*D)$ by remark 6.2.7. Because ∂_i act on the twisted module $\mathcal{O}_V(*D)e^{-f}$ by $\partial_i - \partial_i(f)$, the equation (6.3) transforms in $(i_h)_+\mathcal{O}_V(*D)e^{-f}$ to

$$(t \cdot h^*(\partial_i + a_i) - b_h(-\partial_t t))\delta(t - h)e^{-f} = 0.$$

Assume there exists a polynomial $B(s)$ with $\deg(B) < \deg(b_h)$ such that

$$(tP - B(-\partial_t t))\delta(t - h)e^{-f} = 0$$

holds for some $P \in \mathcal{D}_V[s]$ in $(i_h)_+\mathcal{O}_V(*D)e^{-f}$, then

$$(tP - B(-\partial_t t))\delta(t - h) = 0$$

would hold in $\mathcal{M}[t^{-1}]$, which contradicts the minimality of b_h. □

Theorem 6.2.9. Let h be an equation of the discriminant D of a regular reductive prehomogeneous vector space (G, V) and $f \in V^* \backslash D^*$, then we have $b_{G_1(log\ D)}(s + 1) = b_h(s)$.

Proof. First we recall that the de Rham complex $\mathrm{DR}_V^\bullet(\mathcal{M})$ of a left $\mathcal{D}_V$−module $\mathcal{M}$ represents the (shifted) derived tensor product $\Omega_V^n \otimes_{\mathcal{D}_V}^L \mathcal{M}[-n]$ (see for example [HT07] lemma 1.5.27 or [Bjö93] proposition 2.2.10). In particular we have :

$$\mathcal{H}^n(DR_V^\bullet(\mathcal{M})) = \Omega_V^n \otimes_{\mathcal{D}_V} \mathcal{M}.$$

The direct image $h_+\mathcal{O}(*D)e^{-f}$ is obtained from $(i_h)_+\mathcal{O}_V(*D)e^{-f}$ as the relative de Rham complex of the projection $p\colon T\times V\to T$, i.e. we have

$$\mathcal{H}^i(h_+\mathcal{O}(*D)e^{-f}) = \mathcal{H}^i(p_*\mathrm{DR}^{n+\bullet}_{T\times V/T}((i_h)_*\mathcal{O}_V(*D)e^{-f}[\partial_t])).$$

Using $\Omega^k_{T\times V/T} = \mathcal{O}_T\boxtimes\Omega^k_V$ we get by a direct calculation

$$\mathrm{DR}^\bullet_{T\times V/T}((i_h)_*\mathcal{M}) = (i_h)_*\mathrm{DR}^\bullet_V(\mathcal{M}).$$

Hence we have

$$\mathcal{H}^i(h_+\mathcal{O}_V(*D)e^{-f}) = \mathcal{H}^i(h_*\mathrm{DR}^{n+\bullet}_V(\mathcal{O}_V(*D)e^{-f}[\partial_t])).$$

Because h is affine, h_* is exact and commutes with taking cohomology, so we conclude

$$\mathcal{H}^i(h_+\mathcal{O}_V(*D)e^{-f}) = h_*\mathcal{H}^i(\mathrm{DR}^\bullet_V(\mathcal{O}_V(*D)e^{-f}[\partial_t]))$$

and in particular we have

$$\mathcal{H}^0(h_+\mathcal{O}_V(*D)e^{-f}) = h_*(\Omega^n_V\otimes_{\mathcal{D}_V}\mathcal{O}_V(*D)e^{-f}[\partial_t]).$$

So taking global sections and using lemma 6.2.2, we obtain

$$G_1(*D)\cong \mathrm{H}^0(V,\Omega^n_V\otimes_{\mathcal{D}_V}(\mathcal{O}_V(*D)e^{-f}[\partial_t],$$

where the generator $\omega_1/n_h = vol/dh$ is mapped under this isomorphism to the generator $vol\otimes\delta(t-h)e^{-f}$. By lemma 6.2.8 we have

$$(t\cdot h^*(\partial_i+a_i) - b_h(t\partial_t))\delta(t-h)e^{-f} = 0$$

in $\mathcal{O}_V(*D)e^{-f}[\partial_t]$ and therefore we have

$$vol\otimes(t\cdot h^*(\partial_i+a_i) - b_h(t\partial_t))\delta(t-h)e^{-f} = 0$$

in $h_*(\Omega^n_V\otimes_{\mathcal{D}_V}(\mathcal{O}_V(*D)e^{-f}[\partial_t]))$ and hence

$$t\cdot h^*(\partial_i+a_i)(vol)\otimes\delta(t-h)e^{-f} = b_h(\partial_t t)(vol\otimes\delta(t-h)e^{-f}).$$

The operator $h^*(\partial_i+a_i)$ acts on vol by the right $\mathcal{D}_V-$action on Ω^n_V. So if we write $h^*(\partial_i+a_i) = \sum_{1\le|I|\le n} h_I\partial_1^{i_1}\dots\partial_n^{i_n} + h^*(a_i)$, we get

$$h^*(\partial_i+a_i)(vol) = \sum_{1\le|I|\le n}(-1)^{|I|}h_I(Lie^{i_1}_{\partial_1}\dots Lie^{i_n}_{\partial_n})(vol) + h^*(a_i)(vol).$$

But $Lie_{\partial_i}(vol) = 0$ for all $i = 1,\dots n$, so we conclude that $th^*(a_i) - b_h(-\partial_t t)$ annihilates the section $vol\otimes\delta(t-h)e^{-f}$. It follows that $b_h(-\partial_t t)$ sends $U_0G_1(*D)$ to $U_{-1}G_1(*D)$, i.e. $b_h(s)$ is an element of the ideal $(b_{G_1(log\ D)})$. But as both polynomials are unitary and have degree equal to n_h, they are equal. □

Corollary 6.2.10. Let h be an equation of the discriminant in a regular reductive prehomogeneous vector space. We denote by $\nu_1, \ldots, \nu_{\deg(h)}$ the entries of the matrix A_∞ as defined the beginning of this section. Then the roots α_i of the Bernstein-Sato polynomial of h are given by

$$\alpha_i = \frac{i - 1 - \nu_i}{\deg(h)} - 1.$$

Chapter 7

Quivers and their representation theory

Most of our examples of regular reductive prehomogeneous vector spaces come from the representation theory of quivers. A quiver is a finite directed grap and a representation of a quiver assigns to each node a vector space and to each arrow a linear map. The notion of a quiver representation was introduced in [Gab72] to provide a framework for a wide range of linear algebra problems, for example the classification of tuples of subspaces of a prescribed vector space.
It turned out that quivers and their representations play an important role in the representation theory of finite dimensional algebras and we will explain this in subsection 7.1.
On the other hand looking at quivers from an geometric point of view in subsection 7.2 we see that space V of quiver representation with prescribed dimension vector carries a natural action of an algebraic group G and two representations are isomorphic, if and only if they lie in the same orbit of this group. We will give sufficient and necessary conditions for such a pair (G, V) to be a regular reductive prehomogeneous vector space.
In subsection 7.3 we explain a method due to Schofield to determine the equations of the irreducible components of the discriminant. This method was also independently discovered by [DZ01].
At last in subsection 7.4 we carry out this method in an example. Moreover we will see how in special situations the irreducible component can be read off the quiver directly.
We remark that all the material beside the propositions for reading of the defining equations directly is well known and we refer to [Bri12] and [ARS95] for more detailed introduction to the subject.

We also point out the article [BM06]. Although in [loc.cit.] the authors were interested in the construction of linear free divisors and we work (to some extend) in more general situation, [loc.cit.] was an important source for this thesis.

As we are again in a more general situation, we denote by $\mathbb{F}$ an algebraic closed field and specialize to the case $\mathbb{F} = \mathbb{C}$ in the examples later.

7.1 Quiver representations - an algebraic point of view

In this section we will recall the notion of a quiver, the category of quiver representations and the equivalence of categories between the category of quiver representations and the category of left modules over the path algebra. The equivalence will allow us to apply the representation theory of Artin algebras and to derive Ringel's exact sequence at the end of this subsection.
This exact sequence will play different important roles in our applications, i.e. we derive from the sequence a criterion for prehomogeneity, a way to compute the logarithmic vector fields along the discriminant and to compute the equations of the irreducible components of the discriminant.

Definition 7.1.1. A *quiver* $Q = (Q_0, Q_1, t, h)$ is an oriented graph, i.e. it consists of a set Q_0 of nodes, a set of arrows Q_1 and two maps $t, h\colon Q_1 \to Q_1$ assigning to an arrow α its tail $t\alpha$ and its head $h\alpha$.
We simplify our notation and only write $Q = (Q_0, Q_1)$.

Remark 7.1.2. Additionally we will assume unless something else is said that

- a quiver $Q = (Q_0, Q_1)$ is a connected graph,
- the sets Q_0 and Q_1 are finite and
- Q does not contain oriented cycles.

An *oriented cycle* is a non empty set of arrows $\{\alpha_1, \dots, \alpha_n\}$, such that $t\alpha_{i+1} = h\alpha_i$ for $1 \leq i < n$ and $h\alpha_n = t\alpha_1$. We remark that a cycle with $t\alpha_1 = h\alpha_1$ is also called a loop.

Definition 7.1.3. A *quiver representation* $M = (V_i, V(\alpha))$ of a quiver $Q = (Q_0, Q_1)$ consists of a family of $\mathbb{F}-$vector spaces V_i for all $i \in Q_0$ and a family of $\mathbb{F}-$linear maps $V(\alpha)\colon V_{t\alpha} \to V_{h\alpha}$ for all $\alpha \in Q_1$.
A representation $M = (V_i, V(\alpha))$ is called *finite dimensional*, if each V_i is a finite dimensional vector space. We will always assume that a quiver representation M is finite dimensional and call the tuple

$$\mathbf{d} = (\dim(V_i))_{i \in Q_0}$$

the *dimension vector of* M.

Definition 7.1.4. A *morphism* $\psi\colon M \to N$ *of two representations* is a family of $\mathbb{F}-$linear maps $\psi_i\colon V_i \to W_i$, $i \in Q_0$, such that for all arrows $\alpha \in Q_1$ the following diagram commutes

$$\begin{array}{ccc} V_{t\alpha} & \xrightarrow{V(\alpha)} & V_{h\alpha} \\ \downarrow{\scriptstyle \psi_{t\alpha}} & & \downarrow{\scriptstyle \psi_{h\alpha}} \\ W_{t\alpha} & \xrightarrow{W(\alpha)} & W_{h\alpha}. \end{array}$$

We denote by $\operatorname{Hom}_Q(M,N)$ the $\mathbb{F}-$vector space of all morphisms of representations from M to N and denote it in the special case $M = N$ by $\operatorname{End}_Q(M)$.
For two morphisms $\psi\colon M \to N$ and $\phi\colon N \to L$ of quiver representations, the family of compositions

$$(\phi_i \circ \psi_i)\colon M_i \to L_i,\ i \in Q_0,$$

gives a morphism of representations $\phi \circ \psi\colon M \to L$. Obviously the so-defined composition of morphisms is associative and each representation M has an identity morphism.
We denote the so-defined *category of representations of* Q by $\operatorname{Rep}(Q)$.
We will now describe the category $\operatorname{Rep}(Q)$ in more algebraic terms to derive some of its basic properties and to apply the representation theory of Artin algebras.
Therefore we associate to any representation $M = (V_i, V(\alpha))$ of a fixed quiver $Q = (Q_0, Q_1)$ the vector space

$$V := \bigoplus_{i \in Q_0} V_i$$

equipped with the following two families of endomorphisms:
For all $i \in Q_0$ the maps

$$f_i\colon V \overset{\pi_i}{\to} V_i \overset{\iota_i}{\to} V,$$

which are defined by the composition of the projections π_i and the inclusions ι_i, and for all $\alpha \in Q_1$ the maps

$$f_\alpha : V \to V,$$

which are obtained in a similar way by $V \stackrel{\pi_{t\alpha}}{\to} V_{t\alpha} \stackrel{V(\alpha)}{\to} V_{h\alpha} \stackrel{\iota_{h\alpha}}{\to} V$.
We note that these maps satisfy the relations

$$f_i^2 = f_i,\ f_i f_j = 0\ (i \neq j),\ f_\alpha f_{t\alpha} = f_\alpha = f_{h\alpha} f_\alpha$$

and all other products are zero. This motivates the following:

Definition 7.1.5. The *path algebra* $\mathbb{F}Q$ of a quiver Q is the associative algebra determined by the generators e_i, $i \in Q_0$, and e_α, $\alpha \in Q_1$, and the relations

$$e_i^2 = e_i,\ e_i e_j = 0\ (i \neq j),\ e_{h\alpha}\alpha = \alpha e_{t\alpha} = \alpha. \tag{7.1}$$

So any representation M of Q gives rise to a $\mathbb{F}Q$−module V by $e_i \mapsto f_i$ and $e_\alpha \mapsto f_\alpha$. Moreover this construction extends to functors and yields the following equivalence of categories:

Theorem 7.1.6. ([ARS95] Theorem. III.1.5) The category Rep(Q) is equivalent to the category of (left) finite dimensional $\mathbb{F}Q$−modules.

Remark 7.1.7. (cf. [ARS95] II.1) Let us recall some well known facts of the path algebra of an arbitrary quiver Q, i.e. Q may be not connected, have an infinite number of nodes or contain an oriented cycle.

The path algebra $\mathbb{F}Q$ has an unit element 1 if and only if Q has only a finite number of nodes. In this case

$$1 = \sum_{i \in Q_0} e_i$$

is a decomposition of 1 into orthogonal idempotents.
The Algebra $\mathbb{F}Q$ is a finite dimensional $\mathbb{F}$ vector space if and only if Q has no oriented cycles. In particular $\mathbb{F}Q$ is an Artin algebra, if Q has no oriented cycles.

Hence in our setting theorem 7.1.6 allows us to apply module theory of Artin algebras and we will freely identify quiver representations of Q with $\mathbb{F}Q$−left modules and vice versa in the following. Moreover the equivalence in theorem 7.1.6 is a 1 to 1 correspondences between the projective (resp. injective, indecomposable,...) representations and projective (resp. injective, indecomposable,...) modules and we refer to [ARS95] II for the details.

We recall that a module $M \neq 0$ is called *decomposable*, if M can be written as a direct sum of two non-zero submodules $M = M_1 \oplus M_2$. Otherwise M is called *indecomposable*.

For a quiver representation the definitions are precisely the same. Here the direct sum of two representations $M = (V_i, V(\alpha))$ and $N = (W_i, W(\alpha))$ is the representation $M \oplus N = (U_i, U(\alpha))$, where the $\mathbb{F}-$vector spaces are $U_i = V_i \oplus W_i$, $i \in Q_0$, and the linear maps are

$$U(\alpha) = \begin{pmatrix} V(\alpha) & 0 \\ 0 & W(\alpha) \end{pmatrix}, \ \alpha \in Q_1.$$

The indecomposable modules are of special interest, because we have by the classical *Krull-Schmidt theorem* (e.g. [Lan02] X theorem 7.5):

Theorem 7.1.8. Every finite dimensional $\mathbb{F}Q-$module $M \neq 0$ decomposes into

$$M \cong \oplus_{i=1}^{r} M_i^{m_i},$$

where $M_1, \ldots, M_r$ are indecomposable and pairwise non-isomorphic $\mathbb{F}Q-$modules and the $m_1, \ldots, m_r$ are positive inters. Moreover the indecomposable sumands M_i and their multiplicities m_i are uniquely determined up to reordering.

The indecomposability of a $\mathbb{F}Q-$module can be characterized by its endomorphism ring as follows and a geometric interpretation of this fact will be important later:

Lemma 7.1.9. (cf. [Bri12] Lemma. 1.3.3) A finite dimensional $\mathbb{F}Q-$module M is indecomposable if and only if $\mathrm{End}_{\mathbb{F}Q}(M) = I \oplus \mathbb{F}\mathrm{id}_M$, where $I \subset \mathbb{F}Q$ is a nilpotent ideal.

Obviously by theorem 7.1.6 the category $\mathrm{Rep}(Q)$ is abelian. Moreover, because $\mathbb{F}Q$ is Artin by our assumptions, the category $\mathrm{Rep}(Q)$ is *hereditary*, i.e. every subobject of a projective object is projective.

Indeed by [ARS95] Cor. I 5.2 hereditary is equivalent to the statement that every $\mathbb{F}Q-$module M has a projective resolution of length at most 1 and such a projective resolution of M is given as follows: Because $1 = \sum_{i \in Q_0} e_i$ is a decomposition into orthogonal idempotents, the left $\mathbb{F}Q-$modules $P(i) := \mathbb{F}Qe_i$ are projective indecomposable ([Bri12] Prop. 1.3.7) and every $\mathbb{F}Q-$module M has *the standard (projective) resolution* ([Bri12] prop. 1.4.1):

$$0 \to \bigoplus_{\alpha \in Q_1} P(t\alpha) \otimes_{\mathbb{F}} e_{s\alpha} M \xrightarrow{u} \bigoplus_{i \in Q_0} P(i) \otimes_{\mathbb{F}} e_i M \xrightarrow{v} M \to 0.$$

Here the maps are $u(a\otimes m) := a\alpha\otimes m - a\otimes\alpha m$ and $v(a\otimes m) := am$ and the $\mathbb{F}Q-$module structure on each $P(j)\otimes e_i M$ is given by $a(b\otimes m) = ab\otimes m$.
By [ARS95] Cor. I.5.2 being hereditary can also be characterized by the vanishing of higher Ext−groups. Precisely, for all $i \geq 2$ we have

$$\mathrm{Ext}^i_{\mathbb{F}Q}(M,N) = 0, \text{ for all } M,N \in \mathrm{Obj}(\mathrm{Rep}(Q)).$$

As the groups $\mathrm{Ext}^i(M,N)$ vanish for $i \geq 2$, we write $\mathrm{Ext}_{\mathbb{F}Q}(M,N)$ instead of $\mathrm{Ext}^1_{\mathbb{F}Q}(M,N)$.
If we apply the functor $\mathrm{Hom}_Q(-,N)$ to the standard resolution and use the isomorphism $\mathrm{Hom}_{\mathbb{F}Q}(P(e),M) \cong eM$, given by $f \mapsto f(e)$ ([Ben91] lem. 1.3.3), we obtain *Ringel's exact sequence* ([Rin76] 2.1):

$$0 \longrightarrow \mathrm{Hom}_{\mathbb{F}Q}(M,N) \longrightarrow \bigoplus_{i\in Q_0} \mathrm{Hom}_{\mathbb{F}}(V_i,W_i) \xrightarrow{c_{M,N}} \bigoplus_{a\in Q_1} \mathrm{Hom}_{\mathbb{F}}(V_{t\alpha},W_{h\alpha}) \longrightarrow \mathrm{Ext}_{\mathbb{F}Q}(M,N) \longrightarrow 0$$

Here $M = (V_i, V(\alpha))$ and $N = (W_i, W(\alpha))$ are two $Q-$representations and the map $c_{M,N}$ is defined by

$$(\phi_i)_{i\in Q_0} \mapsto (\phi_{h\alpha}V(\alpha) - W(\alpha)\phi_{t\alpha})_{\alpha\in Q_1}.$$

In the special case $M = N$ we obtain in the category $\mathrm{Rep}(Q)$:

Corollary 7.1.10. (cf. [Bri12] Corollary 1.4.2) Let $M = (V_i, V(\alpha))$ be a $Q-$representation. Then we have the following four term exact sequence

$$0 \longrightarrow \mathrm{End}_Q(M) \longrightarrow \bigoplus_{i\in Q_0} \mathrm{End}(V_i) \xrightarrow{c_M} \bigoplus_{a\in Q_1} \mathrm{Hom}(V_{t\alpha},V_{h\alpha}) \longrightarrow \mathrm{Ext}_Q(M) \longrightarrow 0$$

where $\mathrm{Ext}_Q(M) = \mathrm{Ext}_Q(M,M)$ denotes the *space of self extensions of* M and $c_M = c_{M,M}$ is the map

$$(\phi_i)_{i\in Q_0} \mapsto (\phi_{h\alpha}V(\alpha) - V(\alpha)\phi_{t\alpha})_{\alpha\in Q_1}.$$

In [BM06] section 5 this sequence was interpreted in terms of deformation theory and it was explained, how linear free divisors, a special

class of prehomogeneous vector spaces, appear naturally in the context of quiver representations. Moreover the sequence is used, following [Sch91], to compute semi-invariants of the corresponding prehomogeneous vector space and find explicit equations for the discriminant.
We will explain in the next section, how (regular reductive) prehomogeneous vector spaces arise naturally from the represenation theory of quivers and we will explain the method of [Sch91] in the sections 7.3 and 7.4.

7.2 Quiver representations - a geometric point of view

In this subsection we consider quiver representation from a geometric point of view and see, how quivers gives rise to regular reductive prehomogeneous vector spaces. We derive two criteria for prehmogeneity (cf. rem. 7.2.9 and cor. 7.2.19). We also mention the special class of Dynkin quiver, where the prehomogeneity follows from a simple numerical condition on the Tits form. In the end we give a geometric interpretation of the four-term exact sequence from corollary 7.1.10, which also gives a criterion for the regularity of the prehomogeneous vector space.

Let $(Q = (Q_0, Q_1)$ be a quiver and $M = (V_i, V(\alpha))$ be a quiver representation. If we fix a dimension vector $\mathbf{d} \in \mathbb{N}^{Q_0}$ and bases of the vector spaces V_i, then we can consider M as a point in *the representation space*

$$\operatorname{Rep}(Q, \mathbf{d}) := \bigoplus_{\alpha \in Q_1} \operatorname{Hom}(V_{t\alpha}, V_{h\alpha}) \cong \bigoplus_{\alpha \in Q_1} \operatorname{Mat}_{\mathbf{d}_{h\alpha} \times \mathbf{d}_{t\alpha}}(\mathbb{F}).$$

The group $\operatorname{GL}(Q, \mathbf{d}) := \prod_{i \in Q_0} \operatorname{GL}_{\mathbf{d}_i}(\mathbb{F})$ acts on $\operatorname{Rep}(Q, \mathbf{d})$ by

$$(g_i)_{i \in Q_0} . (V(\alpha))_{\alpha \in Q_1} = (g_{h\alpha} \circ V(\alpha) \circ g_{t\alpha}^{-1})_{\alpha \in Q_1}.$$

Conversely any point $x \in \operatorname{Rep}(Q, \mathbf{d})$ defines a representation M_x of Q and by the assignment $x \mapsto M_x$ we may speak of the action of the group $\operatorname{GL}(Q, \mathbf{d})$ on quiver representations. We conclude from the definition:

Lemma 7.2.1. The $\operatorname{GL}(Q, \mathbf{d})$–orbits in $\mathbf{Rep}(Q, \mathbf{d})$ corresponds to the isomorphism classes of representations of Q with dimension vector $\mathbf{d}$ and the isotropy group $\operatorname{GL}(Q, \mathbf{d})_M$ is isomorphic to the automorphism group $\operatorname{Aut}_Q(M)$.

We remark that the central subgroup

$$\mathbb{F}^* \mathrm{id} = \{(\lambda \cdot \mathrm{id}_{\mathbf{d}_i})_{i \in Q_0} \mid \lambda \in \mathbb{F}^*\}$$

acts trivially on $\mathrm{Rep}(Q,\mathbf{d})$. Hence the $\mathrm{GL}(Q,\mathbf{d})-$action factors through an action of the quotient group

$$\mathrm{PGL}(Q,\mathbf{d}) := \mathrm{GL}(Q,\mathbf{d})/\mathbb{F}^*\mathrm{id},$$

which is an irreducible (hence connected) linear algebraic group of dimension

$$\dim(\mathrm{PGL}(Q,\mathbf{d})) = \dim(\mathrm{GL}(Q,\mathbf{d})) - 1 = \sum_{i\in Q_0} \mathbf{d}_i^2 - 1.$$

Obviously the pair $(\mathrm{GL}(Q,\mathbf{d}), \mathrm{Rep}(Q,\mathbf{d}))$ is a prehomogeneous vector space if and only if $(\mathrm{PGL}(Q,\mathbf{d}), \mathrm{Rep}(Q,\mathbf{d}))$ is a prehomogeneous vector space.
Moreover the groups $\mathrm{GL}(Q,\mathbf{d})$ and $\mathrm{PGL}(Q,\mathbf{d})$ are reductive over $\mathbb{C}$, which is the situation in our applications. Hence it remains to give conditions for the prehomogeneity.
We recall that $\dim(G) \geq \dim(V)$ was a necessary condition for a pair (G,V) to be a prehomogeneous vector space. This condition can reformulated for the pair $(\mathrm{PGL}(Q,\mathbf{d}), \mathrm{Rep}(Q,\mathbf{d})$ by using the Tits form:

Definition 7.2.2. Let $\mathbf{d},\mathbf{e} \in \mathbb{N}^{Q_0}$ be two dimension vectors for a quiver $Q = (Q_0, Q_1)$. The *Euler form* is defined by

$$\langle \mathbf{e},\mathbf{d}\rangle_Q := \sum_{i\in Q_0} \mathbf{e}_i\mathbf{d}_i - \sum_{\alpha\in Q_1} \mathbf{e}_{t\alpha}\mathbf{d}_{h\alpha}.$$

and the *Tits form* is the associated quadratic form $q_Q(\mathbf{d}) = \langle \mathbf{d},\mathbf{d}\rangle_Q$. For representations $M \in \mathrm{Rep}(Q,\mathbf{d})$ and $N \in \mathrm{Rep}(Q,\mathbf{e})$ we also write $\langle M,N\rangle_Q$ for $\langle \mathbf{d},\mathbf{e}\rangle_Q$.

Remark 7.2.3. The following description will be useful later. We choose an ordering of the nodes in Q_0 and think of $\mathbf{e}$ and $\mathbf{d}$ as row vectors and write the Euler form as $\langle \mathbf{e},\mathbf{d}\rangle_Q = \mathbf{e}E\mathbf{d}^T$, where the entries of the *Euler matrix* E are given by

$$E_{i,j} = \delta_{i,j} - \#\{\alpha \in Q_1 \mid t\alpha = i,\ h\alpha = j\},$$

and $\delta_{i,j}$ denotes the Kronecker delta. In this situation the Tits form is associated to the *Cartan matrix* $C = E + E^T$ of Q.
We observe that the Tits form only depends on the underlying undirected graph of Q and determines the undirected graph uniquely.

Remark 7.2.4. The condition $q_Q(\mathbf{d}) \geq 1$ is necessary for the prehomogeneity of $(\mathrm{PGL}(Q,\mathbf{d}), \mathrm{Rep}(Q,\mathbf{d}))$, because $q_Q(\mathbf{d}) = \dim(\mathrm{GL}(Q,\mathbf{d})) - \dim(\mathrm{Rep}(Q,\mathbf{d}))$.

Lemma 7.2.5. Let $\mathbf{d} \in \mathbb{N}^{Q_0}$ be a dimension vector with $q_Q(\mathbf{d}) = 1$. If the orbit $\mathrm{GL}(Q, \mathbf{d}).M$ of $M \in \mathrm{Rep}(Q, \mathbf{d})$ is open, then M is indecomposable.

Proof. Suppose M is decomposable and $M = M_1 \oplus M_2$ is a non trivial decomposition with $M_i \in \mathrm{Rep}(Q, \mathbf{d}_i)$. Then for each M_i there is a one dimensional subgroup $Z_i \cong \mathbb{F}^* \mathrm{id}_{\mathbf{d}_i}$ in the center of $\mathrm{GL}(Q, \mathbf{d}_i)$ acting trivial. As both groups Z_i give rise to different subgroups in the isotropy group $\mathrm{GL}(Q, \mathbf{d})_M$ of M, the isotropy group $\mathrm{GL}(Q, \mathbf{d})_M$ is at least two dimensional. It follows that the $\mathrm{GL}(Q, \mathbf{d})-$orbit of M can not be dense. Indeed because of the assumption $q_Q(\mathbf{d}) = 1$, we have $\dim(\mathrm{Rep}(Q, \mathbf{d})) = \dim(\mathrm{GL}(Q, \mathbf{d})) - 1$. But the dimension of the orbit $\mathrm{GL}(Q, \mathbf{d}).M \cong \mathrm{GL}(Q, \mathbf{d})/\mathrm{GL}(Q, \mathbf{d})_M$ is strictly less than the dimension of the representation space $\mathrm{Rep}(Q, \mathbf{d})$, hence the orbit of M can not be dense. □

We will see in example 7.2.12 that the indecomposability of a representation $M \in \mathrm{Rep}(Q, \mathbf{d})$ is not sufficient for the orbit $\mathrm{GL}(Q, \mathbf{d}).M$ to be open.
Although this can be elaborated directly in example 7.2.12, we give a necessary and sufficient criterion for the prehomogeneity first. Therefore we translate the criterion for the indecomposability of a $\mathbb{F}Q-$module in lemma 7.1.9 to one for the indecomposability of a representation in terms of its automorphism group:

Lemma 7.2.6. ([Bri12] Corollary 2.2.2) A representation $M \in \mathrm{Rep}(Q, \mathbf{d})$ is indecomposable if and only if $\mathrm{GL}(Q, \mathbf{d})_M$ is the semi-direct product of an unipotent subgroup with the group $\mathbb{F}^* \mathrm{id}_{\mathbf{d}}$; equivalently the group $\mathrm{PGL}(Q, \mathbf{d})_M$ is unipotent.

Definition 7.2.7. A representation $M \in \mathrm{Rep}(Q, \mathbf{d})$ is called a *brick* (or *Schur representation*), if $\mathrm{End}_Q(M) = \mathbb{F}\mathrm{id}_{\mathbf{d}}$.

Proposition 7.2.8. ([KR86] Thm. 2.6) A representation $M \in \mathrm{Rep}(Q, \mathbf{d})$ is a brick if and only if M is *stable indecomposable*, i.e. there exists an open neighbourhood of M consisting of indecomposable representations.

Remark 7.2.9. Because the group $\mathrm{Aut}_Q(M)$ is affine and open in its Lie algebra $\mathrm{End}_Q(M)$, the condition $\mathrm{End}_Q(M) = \mathbb{F}_{\mathbf{d}}$ is equivalent to the condition

$$\mathrm{GL}(Q, \mathbf{d})_M = \mathbb{F}^* \mathrm{id}_{\mathbf{d}}.$$

Hence a brick is an indecomposable representation and it is obvious that the orbit of a brick M is open in $\mathrm{Rep}(Q, \mathbf{d})$. In particular this condition will be our main tool to test prehomogeneity.

Definition 7.2.10. We call a dimension vector $\mathbf{d}$ a *real Schur root*, if $\mathrm{Rep}(Q, \mathbf{d})$ contains a brick.

Corollary 7.2.11. If $\mathbf{d}$ is a real Schur root, then $(\mathrm{Rep}(Q, \mathbf{d}), \mathrm{PGL}(Q, \mathbf{d}))$ is a reductive prehomogeneous vector space.

Example 7.2.12. Let Q be the quiver

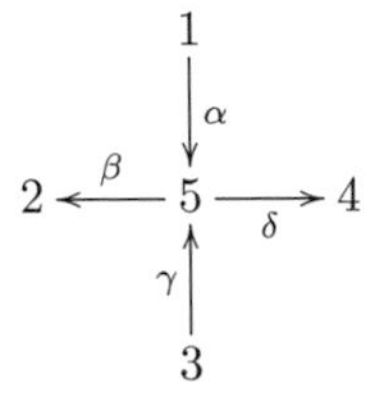

$$\begin{array}{ccccc} & & 1 & & \\ & & \downarrow{\scriptstyle\alpha} & & \\ 2 & \xleftarrow{\beta} & 5 & \xrightarrow[\delta]{} & 4 \\ & & {\scriptstyle\gamma}\uparrow & & \\ & & 3 & & \end{array} .$$

The dimension vector $\mathbf{d} = (1,1,1,1,3)$ fulfils $q_Q(\mathbf{d}) = 1$ and the representation

$$M = (\begin{pmatrix} 1 \\ 0 \\ 0 \end{pmatrix}, \begin{pmatrix} 0 & 1 & 0 \end{pmatrix}, \begin{pmatrix} 0 \\ 0 \\ 1 \end{pmatrix}, \begin{pmatrix} 1 & 1 & 1 \end{pmatrix})$$

is obviously indecomposable, but is not a brick. It follows from a direct calculation that $\dim(\mathrm{GL}(Q, \mathbf{d})_M = 3$. Hence the orbit $\mathrm{GL}(Q, \mathbf{d}).M$ is not open in $\mathrm{Rep}(Q, \mathbf{d})$.

We conclude from example 7.2.12, that in general a indecomposable representation is not a Brick, i.e. not every pair $(Q, \mathbf{d})$ (with $q_Q(\mathbf{d}) = 1$) gives rise to a prehomogeneous vector space $(\mathrm{PGL}(Q, \mathbf{d}), \mathrm{Rep}(Q, \mathbf{d}))$ (with $\dim(\mathrm{PGL}(Q, \mathbf{d})) = \dim(\mathrm{Rep}(Q, \mathbf{d})))$. But there is a special case, where it is:

Suppose the quiver Q is of *finite orbit type*, i.e. there are only finitely many isomorphism classes of representations in $\mathrm{Rep}(Q, \mathbf{d})$ of any prescribed dimension vector $\mathbf{d}$. As the isomorphism classes corresponds to $\mathrm{GL}(Q, \mathbf{d})$−orbits, there are only finitely many $\mathrm{GL}(Q, \mathbf{d})$−orbits in $\mathrm{Rep}(Q, \mathbf{d})$. In this case we conclude from the orbit-lemma, lemma 1.1.3, that one of these orbits is open.
Hence in this case the pair $(\mathrm{GL}(Q, \mathbf{d}), \mathrm{Rep}(Q, \mathbf{d}))$ is a reductive prehomogeneous vector space.
By a theorem of Gabriel all quivers of finite orbit type are classified:

Theorem 7.2.13. (cf. [Gab72] Satz p.3) A (not necessarily connected) quiver Q is of finite orbit type if and only if each connected component of its underlying undirected graph is a simply-laced Dynkin diagram.

The simply-laced Dynkin diagrams are :

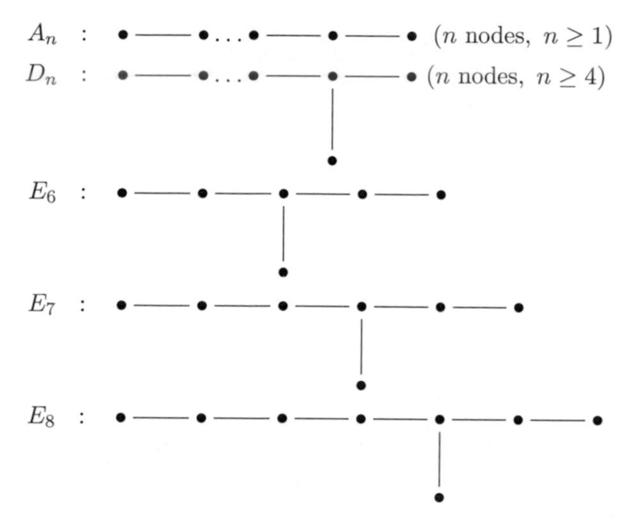

Quivers, whose underlying undirected graph is a disjoint union of simply-laced Dynkin diagrams, are called *Dynkin quivers* and because we are assuming a quiver to be a connected graph, the quivers of finite orbit type for us are precisely the quivers of type A_n, D_n, E_6, E_7 and E_8.

Remark 7.2.14. For completeness we remark the following: By definition an algebra is of *finite orbit type*, if there are only finitely many isomorphism classes of finite dimensional modules of any prescribed dimension. In particular this is by the Krull-Schmidt theorem (theorem 7.1.8) equivalent to the finiteness of isomorphism classes of indecomposable modules of any prescribed dimension. Moreover if there are only finitely many isomorphism classes of indecomposable modules of arbitrary dimension, the algebra is of finite *representation type*.
Obviously finite representation type implies finite orbit type. For a quiver algebra the converse holds and for simplicity we restrict to a connected Dynkin quiver.

Theorem 7.2.15. ([Bri12] Theorem 2.4.3) Let Q be a connected Dynkin quiver, then:

1. Every indecomposable representation is a brick.
2. The dimension vector of the irreducible representations are precisely the $\mathbf{d} \in \mathbb{N}^{Q_0}$ with $q_Q(\mathbf{d}) = 1$.
3. Every irreducible representation is uniquely determined by its dimension vector up to isomorphism.
4. There are only finitely many isomorphism classes of indecomposable representations of Q, i.e. Q is of finite representations type.

Corollary 7.2.16. Let Q be a connected Dynkin quiver and M be a representation with dimension vector $\mathbf{d}$. If $q_Q(\mathbf{d}) = 1$, then the orbit $\mathrm{GL}(Q, \mathbf{d}).M$ is open in $\mathrm{Rep}(Q, \mathbf{d})$.

Remark 7.2.17. Moreover Gabriel showed in [Gab72] that in case of a connected Dynkin quiver Q, the dimension vectors $\mathbf{d}$ with $q_Q(\mathbf{d}) = 1$ are precisely the positive roots of the underlying Dynkin diagram. These positive roots are listed in [Bou81] for example.
We remark that generalization of these results can be found in [Kac80] and [Kac82].

Next we give a geometric interpretation of the Ringel sequence from corollary 7.1.10, which will lead to a second criterion for prehomogeneity.

Theorem 7.2.18. ([Bri12] 2.2.3) Let $M = (V_i, V(\alpha)) \in \mathrm{Rep}(Q, \mathbf{d})$ be a quiver representation.

1. We have an exact sequence

$$0 \to \mathrm{End}_Q(M) \to \bigoplus_{i \in Q_0} \mathrm{End}(V_i) \xrightarrow{c_M} \mathrm{Rep}(Q, \mathbf{d}) \to \mathrm{Ext}^1_Q(M) \to 0,$$

   where $c_M((\phi)_{i \in Q_0}) = (\phi_{h\alpha} M(\alpha) - M(\alpha)\phi_{t\alpha})_{\alpha \in Q_1}$.

2. The map c_M can be identified with the differential at the identity of the orbit map

$$\begin{array}{rcl} \mathrm{GL}(Q, \mathbf{d}) & \to & \mathrm{Rep}(Q, \mathbf{d}) \\ g & \mapsto & g.M. \end{array}$$

3. The image of c_M is the Zariski tangent space $T_M(\mathrm{GL}(Q, \mathbf{d}).M)$ viewed as a subspace of $T_M(\mathrm{Rep}(Q, \mathbf{d}) \cong \mathrm{Rep}(Q, \mathbf{d})$.

This way we obtain a representation theoretic interpretation of the Zariski tangent spaces and their normal spaces to orbits. We recall that for a point M in a locally closed subvariety Y of a variety X the Zariski tangent space $T_M(Y)$ is identified with a subspace of $T_M(X)$ and the *normal space at* M is the quotient $N_M(Y/X) := T_M(X)/T_M(Y)$.

Corollary 7.2.19. ([Bri12] 2.2.5) Let $O_M := \mathrm{GL}(Q, \mathbf{d}).M$ be the orbit of a representation $M \in \mathrm{Rep}(Q, \mathbf{d})$. Then we have isomorphisms

$$\begin{aligned} T_M &\cong \bigoplus_{i \in Q_0} \mathrm{End}(V_i)/\mathrm{End}_Q(M) \\ N_M(O_M/\mathrm{Rep}(Q, \mathbf{d})) &\cong \mathrm{Ext}_Q(M). \end{aligned}$$

Moreover the orbit O_M is open in $\mathrm{Rep}(Q, \mathbf{d})$ if and only if $\mathrm{Ext}_Q(M) = 0$. In this case the orbit O_M is uniquely determined by the dimension vector $\mathbf{d}$.

So far we have only dealt with the question, when a pair $(Q, \mathbf{d})$ gives rise to a reductive prehomogeneous vector space. But for our application concerning the Bernstein-Sato polynomials, we moreover assumed the prehomogeneous vector space to be regular. This condition is by theorem 1.3.2 equivalent to the condition that the discriminant D is a divisor. Using the four term exact sequence from above, [BM06] showed:

Theorem 7.2.20. ([BM06] Proposition 5.1) Let $\mathbf{d}$ be a real Schur root with $q_Q(\mathbf{d}) = 1$. We denote by D the complement of the open orbit, then the following statements hold:

1. D is the set of representations M such that $\mathrm{Ext}(M, M) \neq 0$.
2. D is a divisor in $\mathrm{Rep}(Q, \mathbf{d})$.
3. The image of $c_M : \mathfrak{gl}(Q, \mathbf{d}) \otimes \mathcal{O}_V \to \mathrm{Der}_V$ is contained in $\mathrm{Der}(-log\ D)$.
4. $\Delta = \det(c_M) = 0$ is a defining equation for the discriminant D.

Definition 7.2.21. We call $\Delta = \det(c_M)$ from theorem 7.2.20 the *canonical equation* of the discriminant D.

7.3 Equations for the fundamental semi-invariants

We have seen in the previous sections that if $\mathbf{d}$ is a real Schur root of a quiver Q with $q_Q(\mathbf{d}) = 1$, then the pair

$$(G, V) := (\mathrm{GL}(Q, \mathbf{d}), \mathrm{Rep}(Q, \mathbf{d}))$$

is a reductive prehomogeneous vector space by corollary 7.2.11 and the complement of the open orbit is a divisor $D \subset V$ by theorem 7.2.20. Following [BM06] we will now explain a method to obtain defining equations for the irreducible components of D. This method was independently discovered by [Sch91] and [DZ01].

For a fixed sincere real Schur root $\mathbf{d}$, i.e. $\mathbf{d}_i \neq 0$ for all $i \in Q_0$, we denote by h an (not necessarily reduced) equation for D and by h_i reduced equations of the irreducible components D_i of D. We recall that h and the h_i are semi-invariants for the $\mathrm{GL}(Q, \mathbf{d})-$action on $\mathbb{C}[\mathrm{Rep}(Q, \mathbf{d})]$ and by prop. 1.1.10 the h_i are precisely the fundamental semi-invariants, i.e. they generate of the ring spanned by the semi-invariants.

Let $\mathbf{e}$ be another dimension vector and $M = (V_i, V(\alpha)) \in \mathrm{Rep}(Q, \mathbf{d})$ and $N = (W_i, W(\alpha)) \in \mathrm{Rep}(Q, \mathbf{e})$ be two representations. In the special case $\langle \mathbf{e}, \mathbf{d} \rangle_Q = 0$ the map $c_{N,M}$ of the Ringel sequence

$$0 \longrightarrow \mathrm{Hom}_Q(N, M) \longrightarrow \bigoplus_{i \in Q_0} \mathrm{Hom}(W_i, V_i) \xrightarrow{c_{N,M}} \bigoplus_{a \in Q_1} \mathrm{Hom}(W_{t\alpha}, V_{h\alpha}) \longrightarrow \mathrm{Ext}_Q(N, M) \longrightarrow 0.$$

is given by a square matrix. In this case we define a polynomial map

$$\begin{array}{rcl} c\colon \mathrm{Rep}(Q, \mathbf{e}) \times \mathrm{Rep}(Q, \mathbf{d}) & \to & \mathbb{F} \\ (M, N) & \mapsto & \det(c_{M,N}). \end{array}$$

The map

$$\mathrm{Rep}(Q, \mathbf{e}) \times \mathrm{Rep}(Q, \mathbf{d}) \to \mathrm{Hom}_{\mathbb{F}}\Big(\prod_{i \in Q_0} \mathrm{Hom}_{\mathbb{F}}(\mathbb{F}^{\mathbf{e}_i}, \mathbb{F}^{\mathbf{d}_i}, \prod_{\alpha \in Q_1} \mathrm{Hom}_{\mathbb{F}}(\mathbb{F}^{\mathbf{e}_{t\alpha}}, \mathbb{F}^{\mathbf{d}_{h\alpha}})\Big)$$

sending (N, M) to $c_{N,M}$ is $\mathrm{GL}(Q, \mathbf{e}) \times \mathrm{GL}(Q, \mathbf{d})-$equivariant for the natural actions of this group on the source and the target space. It follows that for a fixed $N \in \mathrm{Rep}(Q, \mathbf{e})$ the restriction $c^N := c_{N,\cdot}$ is a semi-invariant on $\mathrm{Rep}(Q, \mathbf{d})$.

Theorem 7.3.1. ([Sch91] Theorem 4.3) Let Q be a quiver and $\mathbf{d}$ be a sincere real Schur root. The polynomials c^N with $\langle N, M\rangle = 0$ span the ring of semi-invariants $\mathrm{SI}(\mathrm{GL}(Q,\mathbf{d}), \mathrm{Rep}(Q,\mathbf{d})$.

As the characters of $\mathrm{GL}_n(\mathbb{F})$ are just integral powers of the determinant, the characters of $\mathrm{GL}(Q,\mathbf{d})$ are in bijection with the elements of $\mathbb{Z}^{Q_0}$. We will usually identify the weight of a semi-invariant with the image in $\mathbb{Z}^{Q_0}$ of its associated character.
The weights of the semi-invariants c^N and of the canonical equation of the discriminant Δ are determined by $\mathbf{d}$ as follows:
We choose an ordering of Q_0 and consider dimension vectors as row vectors. For a dimension vector $\mathbf{d}$ the *in-degree* $\mathbf{in_d}$ and *out-degree* $\mathbf{out_d}$ are the dimension vectors, whose $i-$th component is defined by:

$$\begin{aligned}\mathbf{in_d}(i) &:= \sum_{\alpha\in Q_1,\ h\alpha=i} \mathbf{d}_{t\alpha}\\ \mathbf{out_d}(i) &:= \sum_{\alpha\in Q_1,\ t\alpha=i} \mathbf{d}_{h\alpha}.\end{aligned}$$

In terms of the Euler matrix E of Q (see remark 7.2.3) the degrees are given by

$$\mathbf{id_d} = \mathbf{d} - \mathbf{d}E,\ \mathbf{out_d} = \mathbf{d} - \mathbf{d}E^T.$$

Lemma 7.3.2. ([BM06] Lemma 6.6) Let $\mathbf{e}, \mathbf{d} \in \mathbb{N}^{Q_0}$ be dimension vectors with $\langle \mathbf{e}, \mathbf{d}\rangle_Q = 0$. Then the weight of the $\mathrm{GL}(Q,\mathbf{d})$ semi-invariant c^N in $\mathbb{Z}^{Q_0}$ is

$$w(c^N) = -\mathbf{e} + \mathbf{in_e} = -\mathbf{e}E$$

and that of the canonical equation Δ is

$$w(\Delta) = \mathbf{in_d} - \mathbf{out_d} = d(E^T - E).$$

It is a well known fact, see for example in [BM06] lemma 3.3. that if Q is a finite quiver without oriented cycles, then its Euler matrix E is invertible.

We will now explain, how we can determine the fundamental semi-invariants in practice for a fixed sincere real Schur root $\mathbf{d}$.
In general we are looking for dimension vectors $\mathbf{e}$ of Q with $\langle \mathbf{e}, \mathbf{d}\rangle_Q = 0$ and compute for a generic $N \in \mathrm{Rep}(Q,\mathbf{e})$ the polynomial c^N.
But if $\ker(c_{N,M}) = \mathrm{Hom}_Q(N, M) \neq 0$, then the matrix $c_{N,M}$ has a non trivial kernel. As c^N is a semi-invariant, the polynomial c^N vanishes on the open orbit $\mathrm{GL}(Q,\mathbf{d}).M$ and by density c^N vanishes on $\mathrm{Rep}(Q,\mathbf{d})$. Hence we obtain:

Lemma 7.3.3. Let $M \in \text{Rep}(Q, \mathbf{d})$ be a brick and and $N \in \text{Rep}(Q, \mathbf{e})$ with $\langle \mathbf{e}, \mathbf{d} \rangle_Q = 0$. Then the polynomial c^N is non-trivial if and only if $\text{Hom}_Q(N, M) = 0$.

Furthermore by the exactness of the Ringel sequence (corollary 7.1.10), any two of the following statements imply the third:

1. $\langle \mathbf{e}, \mathbf{d} \rangle_Q = 0$
2. $\text{Hom}_Q(N, M) = 0$
3. $\text{Ext}^1_Q(N, M) = 0$

Hence we only have to consider generic representations N of the *left orthogonal category* ${}^{\perp}M$, which is by definition the full subcategory, whose objects are those $N \in \text{Obj}(\text{Rep}(Q))$ with

$$\text{Hom}_Q(N, M) = \text{Ext}^1_Q(M, N) = 0.$$

Remark 7.3.4. As the treatment is symmetric, we may also work with the right orthogonal category $M^{\perp}$. These categories are also called *left (resp. right) perpendicular category* and a general study of these categories can be found in [GL91]

Theorem 7.3.5. ([Sch91] Theorem 2.3 or [GL91]) Let Q be a quiver with n nodes and $M \in \text{Rep}(Q, \mathbf{d})$ a brick. Then the left orthogonal category ${}^{\perp}M$ is naturally equivalent to $\text{Rep}(Q')$, where Q' is a quiver with $n-1$ nodes.

Moreover by the following lemma it is enough to consider the simple object of ${}^{\perp}M$:

Lemma 7.3.6. ([DW00] Lemma 1) If

$$0 \to N' \to N \to N'' \to 0$$

is an exact sequence of Q−representations, then we have either

$$\begin{aligned} c^N &= c^{N'} c^{N''} \text{ if } \langle N'', M \rangle_Q = \langle N', M \rangle_Q = 0 \text{ or} \\ c^N &= 0 \text{ if } \langle N', M \rangle_Q < 0. \end{aligned}$$

Indeed if the representation N is not simple, then the polynomial c^N is either zero or a non-trivial product of other polynomials.

Moreover by [Sch91] Theorem 2.5 the category ${}^{\perp}M$ has precisely $n-1 = \#Q_0 - 1$ simple objects N_i and by [Sch91] Theorem 4.3 the semi-invariants c^{N_i} are precisely the fundamental semi-invariants for the $\text{GL}(Q, \mathbf{d})$−action on $\text{Rep}(Q, \mathbf{d})$. We summarize the above discussion:

Corollary 7.3.7. The irreducible components D_i are defined by polynomials c^N, which come from the simple objects in ${}^{\perp}M$.

This also gives an alternative proof of the following theorem, which tells us the number of the fundamental semi-invariants:

Theorem 7.3.8. ([Kac82]) p.153) Let Q be a finite quiver with no oriented cycles and $\mathbf{d}$ be a real Schur root. Then there are precisely $\#Q_0 - 1$ fundamental semi-invariants for the action of $\mathrm{GL}(Q, \mathbf{d})$ on $\mathrm{Rep}(Q, \mathbf{d})$.

Remark 7.3.9. Let us assume, we know the dimension vectors e_i of the $n-1$ simple objects N_i. They form a basis of the free abelian semigroup $\mathbb{N}^{Q'}$ of dimension vectors for ${}^{\perp}M$ and their associated characters $\langle \mathbf{e}_i, \bullet \rangle_Q = w(c^{N_i}) = -\mathbf{e}_i E$ (cf. Lemma 7.3.2) form a basis of the free abelian semigroup of weights for the semi-invariants of the $\mathrm{Rep}(Q, \mathbf{d})-$ action on $\mathrm{Rep}(Q, \mathbf{d})$. Hence knowing the simple objects of ${}^{\perp}M$ we know the weights of the fundamental semi-invariants.
Conversely, knowing the weights w_i of the fundamental semi-invariants h_i, we can calculate the the dimension vectors $\mathbf{e}_i$ by $\mathbf{e}_i = -w_i E^{-1}$.

Remark 7.3.10. In the case of a Dynkin quiver Q the dimension vectors e_i can simply be found by going through the list of positive roots of the Dynkin diagram Q, which are perpendicular to $\mathbf{d}$.
This follows from the facts that by [Sch91] Theorem 2.4 the map $\mathbb{N}^{Q'_0} \to \mathbb{N}^{Q_0}$ sending the $i-$th basis vector to $\mathbf{e}_i$ is an isometry with respect to the Euler forms on Q' and Q, and that simple representations have for Q' have real Schur roots as their dimension vectors.
In general by [ARS95] VIII Prop. 1.7 the dimension vector of any preprojective or preinjective representation M is a real Schur root $\mathbf{d}$, hence there is a huge reservoir of regular reductive prehomogeneous vector spaces and as explained in the proof of [HHKU96] Prop. 2.1 the dimension vectors e_i of the simple objects can be read of the Auslander-Reiten quiver of Q in principle.
However in general it is a difficult problem to to determine the perpendicular category and we refer to [HHKU96] about what is known.

7.4 Example - Finding equations for E_6

In this subsection we demonstrate, how the method explained in 7.3 works in practice. Moreover we prove that in special situations an equation of an irreducible component could be read of the quiver directly. We remark that in [BM06] example 8.1 and section 10 Schofield's method is worked out for the Dynkin quivers E_7 and E_8 with the highest root.

We consider the Dynkin quiver $Q = E_6$ with the following numbering of the nodes and orientation of the arrows:

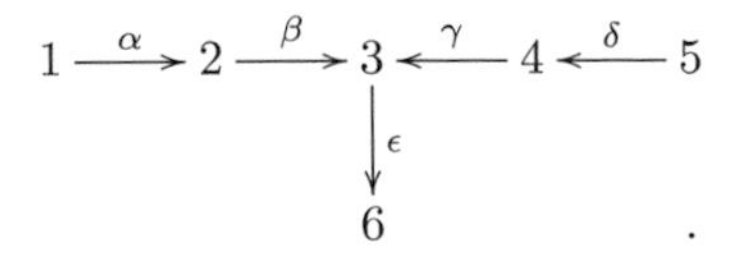

As E_6 is a Dynkin quiver, the dimension vector $\mathbf{d} = (1,1,2,2,1,1)$ is a real Schur root by theorem 7.2.15, because $q_Q(\mathbf{d}) = 1$.
For a general quiver Q we would check that $\mathbf{d}$ is a real Schur root by picking a suitable representation $M \in \mathrm{Rep}(Q,\mathbf{d})$ and either check the condition of remark 7.2.9, i.e. $\mathrm{GL}(Q,\mathbf{d})_M \cong \mathbb{C}^*$, or of corollary 7.2.19, i.e. $\mathrm{Ext}_Q^1(M,M) = 0$.
Before we demonstrate the method of section 7.3, we explain, how we can read off defining equations of the irreducible components of D directly from the quiver in special situations.
We recall that in our situation, i.e. $q_Q(\mathbf{d}) = 1$, the divisor D consists of decomposable representations. Hence natural questions are:

(Q1) When does a representation $M \in \mathrm{Rep}(E_6,\mathbf{d})$:

$$\begin{array}{ccccccccc} \mathbb{C}^{\mathbf{d}_1} & \xrightarrow{A} & \mathbb{C}^{\mathbf{d}_2} & \xrightarrow{B} & \mathbb{C}^{\mathbf{d}_3} & \xleftarrow{C} & \mathbb{C}^{\mathbf{d}_4} & \xleftarrow{D} & \mathbb{C}^{\mathbf{d}_5} \\ & & & & \downarrow E & & & & \\ & & & & \mathbb{C}^{\mathbf{d}_6} & & & & \end{array},$$

decompose?[1]

(Q2) Assume that we found a decomposable representation M and an equation h_i, which vanishes on the orbit $\mathrm{GL}(Q,\mathbf{d}).M$. How do we know that h_i is a defining equation of an irreducible component?

The answer to the second question is easy. By proposition 1.1.10 the defining equations of the irreducible components are irreducible semi-invariants and algebraically independent. Moreover we know that semi-invariants are homogeneous. Summarizing we obtain:

Lemma 7.4.1. Let h_i be be a homogeneous polynomial , which vanishes on the orbit of a decomposable representation. If h_i is an irreducible semi-invariant, then h_i is a defining equation of an irreducible component D_i.

[1] In this section we will also denote the linear maps assigned to the arrows by the capital letters $A,\dots,E$ instead of our standard notation $V(\alpha),\dots,V(\epsilon)$.

Remark 7.4.2. It is convenient use the following convention. We identify $\mathrm{Rep}(Q, \mathbf{d})$ with $\mathbf{C}^n$ by the basis given by the elementary matrices, which we order lexicographically and consider expressions like $\det(A)$ as a function on $\mathrm{Rep}(Q, \mathbf{d})$, where each entry of the matrix is a coordinate function. E.g. in the situation of (Q1) let BA be a square matrix, then the expression $\mathcal{V}(\det(BA))$ would stand for the following vanishing set

$$\mathcal{V}(\det(BA)) = \{(A, B, C, D, E) \in \mathrm{Rep}(Q, \mathbf{d}) \mid \det(BA) = 0\}.$$

Here the entries of the matrices on the right hand side are complex numbers and the entries on the left hand side are coordinate functions.

Before we deal with the first question, we briefly recall the definition of the direct sum in the category of quiver representations :
Let $M' = (V'_i, V'(\alpha))$ and $M'' = (V''_i, V''(\alpha))$ be two representations, then their direct sum $M' \oplus M'' = M = (V_i, V(\alpha))$ is given by

$$\begin{aligned} V_i &= V'_i \oplus V''_i \\ V(\alpha) &= \begin{pmatrix} V'(\alpha) & 0 \\ 0 & V''(\alpha) \end{pmatrix}. \end{aligned}$$

Let us start the discussion in our concrete situation

$$\begin{array}{ccccccccc} 1 & \xrightarrow{\alpha} & 2 & \xrightarrow{\beta} & 3 & \xleftarrow{\gamma} & 4 & \xleftarrow{\delta} & 5 \\ & & & & \downarrow{\scriptstyle \epsilon} & & & & \\ & & & & 6 & & & & \end{array},$$

at the arrow α, but let us consider an arbitrary representation M

$$\begin{array}{ccccccccc} \mathbb{C}^{\mathbf{d}_1} & \xrightarrow{A} & \mathbb{C}^{\mathbf{d}_2} & \xrightarrow{B} & \mathbb{C}^{\mathbf{d}_3} & \xleftarrow{C} & \mathbb{C}^{\mathbf{d}_4} & \xleftarrow{D} & \mathbb{C}^{\mathbf{d}_5} \\ & & & & \downarrow{\scriptstyle E} & & & & \\ & & & & \mathbb{C}^{\mathbf{d}_6} & & & & \end{array},$$

for an arbitrary dimension vector $\mathbf{d} \in \mathbb{N}^6$. Suppose $n_\alpha = \dim(\ker(A)) > 0$. Choosing another representative of the orbit $\mathrm{GL}(Q, \mathbf{d}).M$ means that we have applied a base change to A. Hence we can assume that A is of the form

$$A = \left(\begin{array}{c|c} 0_{n_\alpha} & 0 \\ \hline 0 & \tilde{A} \end{array}\right)$$

and M is isomorphic to a direct sum of the representations

$$M': \qquad \mathbb{C}^{n_\alpha} \xrightarrow{0} 0 \xrightarrow{0} 0 \xleftarrow{0} 0 \xrightarrow{0} 0, \qquad 0 \xrightarrow{0} 0 \text{ (vertical)}$$

$$M'': \qquad \mathbb{C}^{\mathbf{d}_1 - n_\alpha} \xrightarrow{\tilde{A}} \mathbb{C}^{\mathbf{d}_2} \xrightarrow{B} \mathbb{C}^{\mathbf{d}_3} \xleftarrow{C} \mathbb{C}^{\mathbf{d}_4} \xrightarrow{D} \mathbb{C}^{\mathbf{d}_5}, \qquad \mathbb{C}^{\mathbf{d}_3} \xrightarrow{E} \mathbb{C}^{\mathbf{d}_6} \text{ (vertical)}.$$

If $\mathbf{d}_1 = \mathbf{d}_2$, then A is square matrix and the polynomial $\det(A)$ is obviously an irreducible semi-invariant.

Let us generalize this observation. Therefore we denote by $\widetilde{Q_1}$ and $\widetilde{Q_2}$ quivers[2] and by $\mathbb{V}_i$ a representation of $\widetilde{Q_i}$ for $i = 1, 2$. By abuse of language we also allow the case $\widetilde{Q_i} = \emptyset$. Furthermore if the statement is independent of the orientation of an arrow, we replace this arrow by a line.

Lemma 7.4.3. Let

$$M\colon \ \mathbb{V}_1 \longrightarrow \mathbb{C}^{\mathbf{d}_i} \xrightarrow{A} \mathbb{C}^{\mathbf{d}_j} \longrightarrow \mathbb{V}_2$$

be a representation of the quiver

$$Q\colon \ \widetilde{Q_1} \longrightarrow i \xrightarrow{\alpha} j \longrightarrow \widetilde{Q_2}$$

with real Schur root $\mathbf{d}$, such that $\mathbf{d}_i = \mathbf{d}_j$.
If $\dim(\ker(A)) > 0$, then M is isomorphic to the direct sum of the representations

$$M': \qquad \mathbb{V}_1 \longrightarrow V_i' \xrightarrow{0} 0 \longrightarrow 0$$

$$M'': \qquad 0 \longrightarrow V_i'' \xrightarrow{A} \mathbb{C}^{\mathbf{d}_j} \longrightarrow \mathbb{V}_2$$

where $V_i' = \ker(A)$ and $V_i'' = \mathbb{C}^{\mathbf{d}_i} / \ker(A)$. Moreover the set

$$D_\alpha := \{M = (V(\beta))_{\beta \in Q_1} \in \operatorname{Rep}(Q, \mathbf{d}) \mid \det(V(\alpha)) = 0\}$$

is an irreducible component with defining equation $h_\alpha := \det(A) = 0$.

[2]By our general assumption the quivers $\widetilde{Q_1}$ and $\widetilde{Q_2}$ are finite, connected and contain no oriented cycles.

Proof. It is clear that $h_\alpha := \det(A)$ is an irreducible homogeneous polynomial, where A denotes the matrix, whose entries are coordinate functions corresponding to α by our convention. It is also clear that D_α is the vanishing set of h_α. Hence the claim follows, if h_α is a semi-invariant.

Therefore we recall the actions of $\mathrm{GL}(Q, \mathbf{d})$ on $\mathrm{Rep}(Q, \mathbf{d})$ and on regular functions. An element $g = (g_i)_{i \in Q_0} \in \mathrm{GL}(Q, \mathbf{d}) = \prod_{i \in Q_0} \mathrm{GL}_{\mathbf{d}_i}(\mathbb{C})$ acts on a representation $M = (V(\beta)) \in \mathrm{Rep}(Q, \mathbf{d}) = \prod_{\beta \in Q_1} \mathrm{Mat}_{\mathbf{d}_{h\alpha} \times \mathbf{d}_{t\alpha}}(\mathbb{C})$ by

$$g.M = (g_{h\beta} V(\beta) g_{t\beta}^{-1})_{\beta \in Q_1}$$

and on a regular function $h(x)$ by

$$g.h(x) = h(g^{-1}.x).$$

Hence for $h_\alpha(x) = \det(A)$ we obtain

$$\begin{aligned} g.h_\alpha(x) &= g.\det(A) \\ &= \det(g_{h\alpha}^{-1} A g_{t\alpha}) \\ &= \frac{\det(g_{t\alpha})}{\det(g_{h\alpha})} \det(A), \end{aligned}$$

i.e. h_α is a semi-invariant of weight $\frac{\det(g_{t\alpha})}{\det(g_{h\alpha})}$. □

Remark 7.4.4. The statement of lemma 7.4.3 can be easily generalized to the situation, where we replace the arrow α by a path, i.e. a sequence of arrows $\alpha_1, \ldots, \alpha_l$ with $h\alpha_i = t\alpha_{i+1}$ for $i = 1, \ldots, l-1$. The conditions on the real Schur root $\mathbf{d}$ can be stated as follows: For the head and the tail of the path we have

$$\mathbf{d}_{t\alpha_1} = \mathbf{d}_{h\alpha_l}$$

and in between there is a representation $M = (V(\beta))_{\beta \in Q_1}$, such that the linear map

$$V(\alpha_l) \circ \ldots \circ V(\alpha_1) \colon \mathbb{C}^{\mathbf{d}_{t\alpha_1}} \to \mathbb{C}^{\mathbf{d}_{h\alpha_l}}$$

is an isomorphism. If these conditions are satisfied, then $h := \det(\prod_{i=1}^{l} V(\alpha_i))$ is an irreducible semi-invariant of weight $\frac{\det(g_{t\alpha_1})}{\det(g_{h\alpha_l})}$. The proof is essentially the same, as the proof of lemma 7.4.3.

Let us give three easy examples for the second condition:

$$(Ex1) \qquad \mathbb{C}^n \xrightarrow{\alpha_1} \mathbb{C}^{n+1} \xrightarrow{\alpha_2} \mathbb{C}^n$$

$$(Ex2) \qquad \mathbb{C}^n \xrightarrow{\alpha_1} \mathbb{C}^{n-1} \xrightarrow{\alpha_2} \mathbb{C}^n$$

$$(Ex3) \qquad \mathbb{C}^n \xrightarrow{\alpha_1} \mathbb{C}^{n+1} \xrightarrow{\alpha_2} \mathbb{C}^{n+1} \xrightarrow{\alpha_3} \mathbb{C}^n$$

Here n is a natural number and we look at the path only and forgetting the rest of the quiver for a moment.

The examples $(Ex1)$ and $(Ex3)$ fulfil the second condition, where in example $(Ex2)$ the second condition is not fulfilled.
Indeed in $(Ex1)$ we could choose for $V(\alpha_1)$ any injective linear map and then take for $V(\alpha_2)$ the projection to the subspace given by the image of $V(\alpha_1)$. Then obviously the composition $V(\alpha_2) \circ V(\alpha_1)$ is the identity on $\mathbb{C}^n$.
In $(Ex3)$ we could choose for $V(\alpha_1)$ any injective linear map, for $V(\alpha_2)$ the identity and then for for $V(\alpha_3)$ the projection to the subspace defined by the image of $V(\alpha_1)$.
On the other hand in $(Ex2)$ the composition of any linear maps $V(\alpha_2)$ and $V(\alpha_1)$ has a non trivial kernel. Hence $V(\alpha_2)V(\alpha_1)$ can not be an isomorphism for any choice of representation.

Let us consider the case next, where we have more than two in- or outgoing arrows at a node i.
We describe the situation as follows: Let Q_l^{in} and Q_j^{out} be subquivers of a quiver Q indexed by the sets

$$L = \{\alpha \in Q_1 \mid h\alpha = i\}$$

and

$$J = \{\beta \in Q_1 \mid t\beta = i\}.$$

We denote by $n_L = \#L$ and $n_J = \#J$ and assume $n_L + n_J \geq 3$. Then we write Q as

$$Q_l^{in} \overset{\alpha_l}{\rightrightarrows} i \overset{\beta_j}{\rightrightarrows} Q_j^{out},$$

where the symbol $Q_l^{in} \overset{\alpha_l}{\rightrightarrows}$ denotes the n_L ingoing arrows $Q_l^{in} \overset{\alpha_l}{\to} i$, $l \in L$ and the symbol $\overset{\beta_j}{\rightrightarrows} Q_j^{out}$ denotes the n_J outgoing arrows $i \overset{\beta_j}{\to} Q_j^{out}$, $j \in J$.
Similar we write a representation M of Q as

$$\mathbb{K}_l \overset{A_l}{\rightrightarrows} V_i \overset{B_j}{\rightrightarrows} \mathbb{I}_j,$$

where $\mathbb{K}_l$ are representations of Q_l^{in} and $\mathbb{I}_j$ are representations of Q_j^{out}.

Lemma 7.4.5. If the images of the maps A_l, $l \in L$, and the kernels of the maps B_j, $j \in J$, does not span V_i, then the representation M decomposes.

Proof. We put $V_i' = \sum_{l\in L} \mathrm{im}(A_k) + \sum_{j\in J} \ker(B_j)$ and $V_i'' = V_i/V_i'$. Hence M is isomorphic to the direct sum of the representations

$$M': \qquad \mathbb{K}_l \overset{A_l}{\rightrightarrows} V_i' \overset{B_j}{\rightrightarrows} \mathbb{I}_j$$

$$M'': \qquad 0_l \overset{0}{\rightrightarrows} V_i'' \overset{0}{\rightrightarrows} 0_j.$$

Here 0_k denotes the zero-representation, i.e. the vector spaces at each node are equal to the zero space and the linear map for each arrow is the zero map. □

Unfortunately in this general situation we do not know a formula for a defining equation, but in the following we do:
For the quiver Q

$$1 \overset{\alpha}{\to} 2 \overset{\beta_n}{\rightrightarrows} 3.$$

the dimension vector $\mathbf{d} = (n, n, 1)$ is a real Schur root as we will see in section 8.6. We consider the representation

$$\mathbb{C}^n \overset{A}{\to} \mathbb{C}^n \overset{B_n}{\rightrightarrows} \mathbb{C}^1.$$

and denote by B be the matrix, whose $i-$th row is given by the matrix B_i for $i = 1, \ldots, n$.

Lemma 7.4.6. In the situation above $h = \det(B)$ is an irreducible semi-invariant.

Proof. Let $g = (G_1, G_2, \lambda) \in \mathrm{GL}(Q, \mathbf{d})$, then by the definition of the $\mathrm{GL}(Q, \mathbf{d})-$action on $\mathbb{C}[\mathrm{Rep}(Q, \mathbf{d})]$, we have

$$g.h = g.\det(B) = \frac{\det(G_2)}{\lambda^{-1}} \det(B).$$

We conclude that h is a semi-invariant of weight $\frac{\det(G_2)}{\lambda^{-1}}$. □

Remark 7.4.7. Lemma 7.4.6 can also be applied in the situations, when all arrows β_j have the same target or the arrow α is replaced by a path as in remark 7.4.4.

Let us come back to our example E_6 with the representation:

$$\begin{array}{ccccccccc} \mathbb{C}^1 & \xrightarrow{A} & \mathbb{C}^1 & \xrightarrow{B} & \mathbb{C}^2 & \xleftarrow{C} & \mathbb{C}^2 & \xleftarrow{D} & \mathbb{C}^1 \\ & & & & \downarrow{\scriptstyle E} & & & & \\ & & & & \mathbb{C}^1 & & & & \end{array}$$

We know by Kac's Result (Theorem 7.3.8) that there are precisely $\#Q_0 - 1 = 5$ fundamental semi-invariants h_i, which are equations of the irreducible components of the discriminant.

By lemma 7.4.3 and remark 7.4.4 we know 4 of the fundamental semi-invariants, namely

$$\begin{aligned} h_1 &= \det(A) \\ h_2 &= \det(C) \\ h_3 &= \det(EB) \text{ and} \\ h_4 &= \det(ECD). \end{aligned}$$

Furthermore the degree of the canonical equation equals the dimension of the representations space, i.e. we have $\deg(\Delta) = \dim(\mathrm{Rep}(Q, d) = 11$. Hence without further informations we would conclude that the degree of the missing equation h_5 is one or three. But E_6 is a Dynkin quiver, hence the discriminant is a linear free divisor by [BM06] corollary 5.5 and Δ is reduced. We conclude that the degree of h_5 is three.

We will now use Schofield's method to determine h_5.

At first we determine the defining equation for the orthogonal dimension vectors $\mathbf{e}$, i.e. the equation $(\mathbf{e}, \mathbf{d}) = 0$:

$$\begin{aligned} (\mathbf{e}, \mathbf{d}) &= \sum_{i \in Q_0} \mathbf{e}_i \mathbf{d}_i - \sum_{\alpha \in Q_1} \mathbf{e}_{s\alpha} \mathbf{d}_{t\alpha} \\ &= 1\mathbf{e}_1 + 1\mathbf{e}_2 + 2\mathbf{e}_3 + 2\mathbf{e}_4 + 1\mathbf{e}_5 + 1\mathbf{e}_6 \\ &- 1\mathbf{e}_1 - 2\mathbf{e}_2 - 2\mathbf{e}_4 - 2\mathbf{e}_5 - \mathbf{e}_3 \\ &= -\mathbf{e}_2 + \mathbf{e}_3 - \mathbf{e}_5 + \mathbf{e}_6 \end{aligned}$$

The dimension vector $\mathbf{e} = (0, 1, 1, 1, 1, 1)$ is a root of this polynomial and a real Schur root. For example a brick is given by $b = c = d = e = 1$. In order to compute the map $c^N = c_{N,\bullet}$ for a generic representation $N \in \mathrm{Rep}(Q, \mathbf{e})$ we consider the diagram

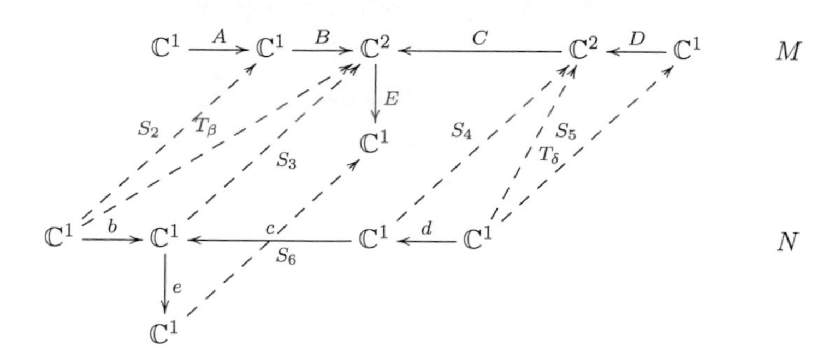

Here the solid arrows indicate maps within the representation, the dotted arrows indicate maps between the representations N and M, i.e. $S_i \in \mathrm{Hom}(\mathbb{C}^{\mathbf{e}_i}, \mathbb{C}^{\mathbf{d}_i})$ for $i \in Q_0$ and $T_\alpha \in \mathrm{Hom}(\mathbb{C}^{\mathbf{e}_{t\alpha}}, \mathbb{C}^{\mathbf{d}_{h\alpha}})$ for $\alpha \in Q_1$. For simplicity we have only drawn the maps T_β and T_δ.
Moreover for each p, q we have ordered the basis of $\mathrm{Hom}(\mathbb{C}^p, \mathbb{C}^q)$ given by the elementar matrices E_{ij}, $1 \leq i \leq q$, $1 \leq j \leq p$, lexicographically. We have

$$c_{N,M} \begin{pmatrix} S_2 \\ S_3 \\ S_4 \\ S_5 \\ S_6 \end{pmatrix} = \begin{pmatrix} S_3 b - B S_2 \\ S_3 c - C S_4 \\ S_4 d - D S_5 \\ S_6 e - E S_3 \end{pmatrix}.$$

Hence $c_{N,M}$ is given by the matrix

$$c_{N,M} = \begin{pmatrix} -B & b\mathrm{I}_2 & 0 & 0 & 0 \\ 0 & c\mathrm{I}_2 & -C & 0 & 0 \\ 0 & 0 & d\mathrm{I}_2 & -D & 0 \\ 0 & -E & 0 & 0 & e\mathrm{I}_1, \end{pmatrix}$$

where the five columns refer to the five maps $S_2, \ldots, S_6$ and the four rows to the four maps $T_\beta, \ldots, T_\epsilon$. Because N is a generic representation, we have $b, c, d, e \neq 0$. Therefore by row operation the matrix transforms to

$$c_{N,M} = \begin{pmatrix} -B & b\mathrm{I}_2 & 0 & 0 & 0 \\ 0 & c\mathrm{I}_2 & -C & 0 & 0 \\ 0 & 0 & d\mathrm{I}_2 & -D & 0 \\ 0 & 0 & 0 & 0 & e \end{pmatrix}.$$

As we have remarked, the representation N given by $b = c = d = e = 1$ is a brick, so we specialize the above matrix to this case and obtain

$$c_{N,M} = \begin{pmatrix} -B & \mathrm{I}_2 & 0 & 0 & 0 \\ 0 & \mathrm{I}_2 & -C & 0 & 0 \\ 0 & 0 & \mathrm{I}_2 & -D & 0 \\ 0 & 0 & 0 & 0 & 1. \end{pmatrix}.$$

Hence (up to a sign) we have $det(c^N) = \det(B|CD)^3$. It is an irreducible polynomial of degree 3 and is a semi-invariant of weight

$$w(c^N) = \begin{pmatrix} 0 & 1 & \text{-1} & 0 & 1 \\ & & 0 & & \end{pmatrix}.$$

This follows from a direct calculation or by lemma 7.3.2. Hence the missing fundamental semi-invariant is $h_5 = \det(B|CD)$.
Of course we can obtain the equations $h_1, \ldots, h_4$ by the same procedure. We summarize the informations in the following table:

Polynomial	Degree	$\text{Root}^{\perp \mathbf{d}}$	Weight
$h_1 = \det(A)$	1	1 0 0 0 0 0	1 -1 0 0 0 0
$h_2 = \det(C)$	2	0 0 0 1 0 0	0 0 -1 1 0 0
$h_3 = \det(EB)$	2	0 1 1 0 0 0	0 1 0 0 0 -1
$h_4 = \det(ECD)$	3	0 0 1 1 1 0	0 0 0 0 1 -1
$h_5 = \det(B\|CD)$	3	0 1 1 1 1 1	0 1 -1 0 1 0
$\Delta = h_1 \ldots h_5$	11		-1 -1 2 -1 -2 2

[3]Here $(B|CD)$ denotes the concatenation of the matrices B and CD.

Chapter 8

Calculations for classical Singularities

In this chapter we present results of the computation of the spectrum and the roots of the Bernstein-Sato polynomial attached to discriminants in quiver representation spaces as described in chapter 7. Therefore we have implemented the methods of the preceding chapters in SINGULAR [DGPS18].
Moreover we introduce in section 8.1 the invariant subspace conjecture 8.1.1 and the exponent condition 8.1.4 and check them for the examples.

Let us briefly recall the setting: Let $(Q, \mathbf{d})$ be a quiver with real Schur root $\mathbf{d}$, such that $q_Q(\mathbf{d}) = 1$. In this case the group $G = \mathrm{GL}(Q, \mathbf{d})$ and the vector space $V = \mathrm{Rep}(Q, \mathbf{d})$ give rise to a regular reductive prehomogeneous vector space (G, V). We denote by h a defining equation of the discriminant D of degree n_h and by $f \in V^* \backslash D^*$ a generic linear function on V. Furthermore we denote by $\mathfrak{g}$ the Lie algebra of G and by $\mathfrak{a} = \ker(d\chi)$, where χ is the character of h.
The following algorithm first computes the matrix of the residue endomorphism of $(G_1(log\ D), \nabla)$ (the restriction of the FL-Brieskorn lattice to $\theta = 1$) in the basis given by the powers $(-f)^i$, for $i = 0, \ldots, n_h - 1 = \deg(h) - 1$. Then it uses the special shape of the relative connections to compute the roots of the Bernstein-Sato polynomial b_h.

"The simple algorithm":

1. Determine a (minimal) set of generators $(\xi_1, \ldots, \xi_{n_h-1})$ of $\mathcal{O}_V \otimes \mathfrak{a}$. E.g. determine generators $(\xi'_1, \ldots, \xi'_{n_h})$ of $\mathrm{im}(c_M : \mathfrak{g} \to \mathrm{Der}(-log\ D))$, such that ξ'_{n_h} is an Euler field, i.e. $\xi'_{n_h}(h) = ch$ for some $c \in \mathbb{C}^*$. Then put $\xi_{n_h} = \frac{\xi'_{n_h}}{c}$ and $\xi_i = \xi'_i - \frac{\xi'_i(h)}{h}\xi_{n_h}$.

2. Define ideals $J_f := df(\xi_1, \ldots, \xi_{n_h-1})$, $Div := J_f + (h)$ and compute a Groebner basis of each.

3. Define $n_h \times n_h$ matrices $\Omega = \begin{pmatrix} 0 & 0 & \ldots & 0 & 0 \\ 1 & 0 & \ldots & 0 & 0 \\ 0 & 1 & \ldots & 0 & 0 \\ \ldots & \ldots & \ldots & \ldots & \ldots \\ 0 & 0 & \ldots & 1 & 0 \end{pmatrix}$

   and $D = \mathrm{diag}(0, 1, \ldots, n_h - 1)$

4. Divide $(-f)^{n_h}$ by h with respect to the ideal Div, i.e.

$$(-f)^{n_h} = \underbrace{\sum_{i=1}^{n_h-1} g_i^{(n_h)} \xi_i(f) + c_{n_h} h}_{=:Div}$$

5. For $j = n_h - 1$ to $j = 1$ do

   (a) $g = \sum_{i=1}^{n_h-1} \xi_i(g_i^{(j+1)})$

   (b) $Div = J_f + (-f)^j$

   (c) Divide g with respect to the ideal Div, i.e. $g = \sum_{i=1}^{n_h-1} g_i^{(j)} \xi_i(f) + c_j(-f)^j$

   (d) $\Omega_{j+1,n_h} = c_j$

   (e) $j = j - 1$

6. Put $\Omega = \frac{1}{n_h}(\Omega - D)$

7. The roots of $b_h(s)$ are the shifted eigenvalues of Ω by -1

We recall that step 5. (a) comes from the application of the division lemma (lemma 5.2.3). Moreover we note that it is similar to the one used in [Sev11] for linear free divisors and compared to the one of [GMS09] it avoids solving the Birkhoff problem for the FL-Brieskorn lattice[1].

However this algorithm does not only run into memory problems during the Groebner basis computations, when the degree of h is rising. Due to memory limitations the basis elements $(-f)^i$ can not even be represented for high i.
Hence we use a different approach. We build a basis in blocks with respect to the irreducible factors of h. This way we lower the degrees of the involved polynomials in the Groebner basis computations.
Let $h = \prod_{j=1}^m h_j$ be a decomposition into irreducible factors and $d_j = \deg(h_j)$. (We recall that we do not require h to be reduced, so a factor h_j may occur more than once in the decomposition.) We consider the following basis $(\omega_0, \ldots, \omega_{n_h-1})$:
For $0 \leq r \leq m-1$ we set

$$\omega_{\sum_{s=1}^r d_s + k_r} = \begin{cases} (-f)^{k_r} \prod_{j=1}^{r} h_j, & \text{for } k_r = 1, \ldots, d_{r+1} - 1 \\ \prod_{j=1}^{r+1} h_j, & \text{for } k_r = d_{r+1} \end{cases},$$

and replace

$$\omega_{n_h} = \omega_{\sum_{s=1}^m d_s} := \omega_0 := 1.$$

Now roughly said, we run the simple algorithm for each irreducible factor h_j. But in this basis we have to consider the characters χ_{h_j} of the irreducible components h_j, when applying the division lemma (lemma 5.2.3), i.e. we replace step 4. and 5. of „The simple algorithm“by the following steps:

[1]It turned out in the computations that solving the Birkhoff problem for the FL-Brieskorn lattice is not much more expensive in terms of time or memory consumption than the first steps 1. to 5. above, which are necessary in any case. However, it is cumbersome to explicitly write down an algorithm for solving a Birkhoff problem, from which we therefore refrain.

„The extended algorithm“

For $r = 1$ to $m-1$ do

(a) Put $Div = J_f + (h_m)$

(b) Divide $(-f)^{\deg(h_r)}$ with remainder with respect to Div, i.e. $(-f)^{\deg(h_r)} = \sum_{i=1}^{n_h-1} g_i\xi_i(f) + c_g h_r$

(c) Put $\Omega_{\sum_{s=1}^{r} d_s, \sum_{s=1}^{r} d_s} = c_g$

(d) Define $g = \sum_{i=1}^{n_h-1} \left(\xi_i(g_i) + \sum_{l=1}^{m-1} g_i\chi_l(\xi_i)\right)$

(e) Run procedure **Div-wrt-f**(polyomial g, integer m):

i. Put $Div = J_f + (-f)^{\deg(g)}$

ii. Divide g with remainder with respect to Div, i.e. $g = \sum_{i=1}^{n_h-1} g_i\xi_i(f) + c_g(-f)^{\deg(g)}$

iii. Put $\Omega_{\sum_{s=1}^{r-1} d_s+\deg(g), \sum_{s=1}^{r} d_s} = c_g$

iv. Put $g := \sum_{i=1}^{n_h-1} \left(\xi_i(g_i) + \sum_{l=1}^{r-1} g_i\chi_l(\xi_i)\right)$

v. If $\deg(g) > 0$ then run **Div-wrt-f(g,r)**

For $r = m$ do

(a) Put $Div = J_f + (h_m)$

(b) Divide $(-f)^{\deg(h_m)}$ with respect to the ideal Div, i.e. $(-f)^{\deg(h_m)} = \sum_{i=1}^{n_h-1} g_i\xi_i(f) + c_g h_m$

(c) Define $g = \sum_{i=1}^{n_h-1} \left(\xi_i(g_i) + \sum_{l=1}^{m-1} g_i\chi_l(\xi_i)\right)$

(d) Run procedure **Div-wrt-f2**(polyomial g):

i. Put $Div = J_f + (-f)^{\deg(g)}$

ii. Divide g with respect to the ideal Div, i.e. $g = \sum_{i=1}^{n_h-1} g_i\xi_i(f) + c_g(-f)^{\deg(g)}$

iii. Put $\Omega_{\sum_{s=1}^{m-1} d_s+\deg(g)+1, n_h} = c_g$

iv. Put $g := \sum_{i=1}^{n_h-1} \left(\xi_i(g_i) + \sum_{l=1}^{m-1} g_i\chi_l(\xi_i)\right)$

v. If $\deg(g) > 0$ then run **Div-wrt-f2(g)**

Remark. We recall our convention: We identify $\text{Rep}(Q, \mathbf{d})$ with $\mathbf{C}^n$ by the basis given by the elementary matrices (ordered lexicographically) and consider expressions like $\det(A)$ as a function on $\text{Rep}(Q, \mathbf{d})$, where each entry of the matrix is a coordinate function.
For two matrices A, B with the same numbers of rows we denote by $(A|B)$ their concatenation, i.e. let $a_1, \ldots, a_l$ (resp. $b_1, \ldots, b_k$) be the columns of A (resp. B), then $(A|B) = (a_1, \ldots, a_l, b_1, \ldots, b_k)$.

8.1 The exponent condition and the invariant subspace conjecture

In this subsection we will discuss the so-called invariant subspace conjecture for a given prehomogeneous action as well as a condition (called exponent condition) on the roots of the Bernstein-Sato polynomial of a defining equation of the discriminant of such an action. Both will later be checked in the examples that we compute.

Let us start with the invariant subspace conjecture. In his Habilitation thesis [Sev09] C. Sevenheck made the following conjecture under the additional assumption that the discriminant D is a linear free divisor.

Invariant subspace conjecture 8.1.1. Let (G, V) be a regular reductive prehomogeneous vector space with $\dim(G) = \dim(V) = n$ and $A_1, \ldots, A_n$ be a basis of the Lie algebra $\mathfrak{g}$. Then the following holds for the equation $h = \det(A_1 x| \ldots |A_n x)$ of the discriminant $D \subset V$:

(C1) There is a maximal torus $T \leq G$ and a $T-$invariant linear sub space $W_T \leq V$, such that $\dim(T) = \dim(W_T)$ and $W_T \not\subset D$.

Suppose (C1) holds, then we have moreover:

(C2) The restriction $h_{|W_T}$ is a monomial.

(C3) The multiplicity of the root -1 of the Bernstein-Sato polynomial b_h of h equals $\dim(T)$.

If D is a linear free divisor, then h is by Saito's criterion (proposition 3.1.15) reduced equation for D.

The interest of this conjecture stems from the relation between Gauss-Manin systems of hyperplane sections of prehomogeneous discriminants (as studied in this work) and of hyperplane sections of Milnor fibres of monomials, where the latter ones are important in toric mirror symmetry (namely, they compute quantum $\mathcal{D}-$modules of weighted projective spaces).

We explain now, how a solution to condition $(C1)$ of the invariant subspace conjecture 8.1.1 is constructed in the examples.

(More precisely, we have two approaches, which are closely connected. We will discuss both approaches in full details later. One approach is discussed in the example D_n and the other one in the example E_6 with the root $\mathbf{d}^1$.)

In both cases we choose a maximal torus $T \leq G$ and an element $v \in V$ of the open $G-$orbit. For the first approach we define the space

$$W_T := \overline{T.v}$$

as the Zariski closure of the T−orbit of v. As the isotropy group G_v is zero dimensional, the isotropy group T_v is zero dimensional too. Hence by

$$\dim(W_T) = \dim(T.v) = \dim(T) - \dim(T_v),$$

we have $\dim(T) = \dim(W_T)$. Moreover as $v \notin D$, also the condition $W_T \not\subset D$ is satisfied. However it is not clear in general, that the closure of the orbit $W_T = \overline{T.v}$ is a $\mathbb{C}$−vector space.
For the second approach we choose a basis $A_1, \ldots, A_m$ of the Lie algebra $\mathfrak{t}$ and define

$$W_T := \langle A_1.v, \ldots, A_m.v \rangle_{\mathbb{C}}.$$

Obviously the space W_T is a vector space and it is T−invariant by [DG70] C.II §6 proposition 2.1, i.e. T and $\mathfrak{t}$ stabilize the same subspaces of V. Hence if we knew that (T, W_T) is a prehomogeneous vector space, we would conclude again $W_T \not\subset D$. Indeed as v is an element of the open orbit, we have in this case by proposition 1.1.4:

$$\dim(W_T) = \dim(\mathfrak{t}) = \dim(T).$$

In all examples it will be easy to verify that the space $\overline{T.v}$ and the space $\langle A_1.v, \ldots, A_n.v \rangle$ coincide and the conditions $(C2)$ and $(C3)$ hold.

Next we explain the connection to GKZ−systems and hypergeometric $\mathcal{D} := \mathbb{C}[t, t^{-1}]\langle \partial_t \rangle$−modules. We begin recalling the definition of an hypergeometric $\mathcal{D}$−module:

Definition 8.1.2. Let n, m be natural numbers, let $\alpha_1, \ldots, \alpha_n, \beta_1, \ldots, \beta_m$ be complex numbers and let $\gamma \in \mathbb{C}^*$. The *one dimensional hypergeometric $\mathcal{D}$−module* $\mathcal{H}_\gamma(\alpha_i, \beta_j)$ is the quotient

$$\mathcal{H}_\gamma(\alpha_i, \beta_j) := \frac{\mathcal{D}}{\gamma \prod_{i=1}^n (t\partial_t - \alpha_i) - t \prod_{j=1}^m (t\partial_t - \beta_j)}.$$

It is well known that a hypergeometric $\mathcal{D}$−module is holonomic. Moreover for $m = n$ the module $\mathcal{H}_\gamma(\alpha_i, \beta_j)$ has regular singularities at the origin, at infinity and at γ. In the case $n > m$ (resp. $n < m$) it has a regular singularity at the origin and an irregular at infinity (resp. regular at infinity and irregular at the origin).
The logarithms of the eigenvalues of the local monodromy (up to a product by $2\pi i$) of $\mathcal{H}_\gamma(\alpha_i, \beta_j)$ are:

$$\begin{aligned} \text{At } 0 &: \alpha_1, \ldots, \alpha_n \\ (\text{When } n = m) \text{ At } \gamma &: 1, \ldots, 1, -\sum \beta_j - \sum \alpha_i \\ \text{At } \infty &: \beta_1, \ldots, \beta_m. \end{aligned}$$

Moreover the isomorphism class of $\mathcal{H}_\gamma(\alpha_i, \beta_j)$ determines n, m and the set of all α_i and β_j modulo the integers. We refer to the book [Kat90] for these statements and more properties of hypergeometric $\mathcal{D}-$modules.

We will now consider the special cases of hypergeometric $\mathcal{D}-$modules, where $m = 0$:

$$\mathcal{H}_\gamma(\alpha_i) = \frac{\mathbb{C}[t, t^{-1}]\langle\partial_t\rangle}{\gamma\prod_{i=1}^n(t\partial_t - \alpha_i) - t}.$$

In particular the restriction $G_1(*D)$ of the FL-Gauss-Manin system to $\theta = 1$ are of that kind. We recall from lemma 6.2.1

$$G_1(*D) \cong \frac{\mathbb{C}[t, t^{-1}]\langle\partial_t\rangle}{(b_{G_1(log\ D)}(t\partial_t) - ct)},$$

where c was a non zero complex number and

$$b_{G_1(log\ D)}(s) = \prod_{i=1}^{n_h}(s - \frac{i - 1 - \nu_i}{n_h})$$

was the spectral polynomial.

Remark 8.1.3. Notice that if $(\alpha_1', \ldots, \alpha_n') \in \mathbb{C}^n$ are such that $\alpha_i' - \alpha_i \in \mathbb{Z}$ for $i = 1, \ldots, n$, then we have $\mathcal{H}_\gamma(\alpha_i') \cong \mathcal{H}_\gamma(\alpha_i)$.

The exponent condition 8.1.4. We say that a hypergeometric $\mathcal{D}-$module $\mathcal{H}(\alpha_i)$ fulfils the *exponent condition*, if there exists positive natural numbers $q_1, \ldots, q_k$ such that $\mathcal{H}(\alpha_i) \cong \mathbb{C}[t, t^{-1}]\langle\partial_t\rangle/(P)$ where

$$P = \gamma(\prod_{i=1}^{k}\prod_{j=0}^{q_i - 1}(t\partial_t - \frac{j}{q_i})) - t. \tag{8.1}$$

In particular this implies $\alpha_i \in \mathbb{Q}$.

Remark 8.1.5. It is clear that we have $\mathcal{D}/(P) = \mathcal{D}/(P')$, where

$$P' = \gamma\prod_{i=1}^{k} q_i^{-q_i}\prod_{j=0}^{q_i - 1}(q_i t\partial_t - j) - t.$$

Next we recall the definition of a *GKZ-system*, which can be considered as a higher-dimensional versions of classical hypergeometric systems in one variable.

Definition 8.1.6. Let $\beta = (\beta_1, \dots, \beta_d) \in \mathbb{C}^d$ and $A = (a_{ki})$ be a $d \times n$ integer matrix the columns of which generate $\mathbb{Z}^d$. We denote by $\mathbb{L}_A$ the $\mathbb{Z}-$module of integer relations among the columns of A, i.e. the kernel of the linear mapping $\mathbb{Z}^n \to \mathbb{Z}^d$ given by A. We define

$$\mathrm{M}_A^\beta := \frac{\mathbb{C}[\lambda_1, \dots, \lambda_n]\langle \partial_{\lambda_1}, \dots, \partial_{\lambda_n} \rangle}{(I)},$$

where I is the left ideal generated by the operators

$$\Box_{\underline{l}} = \prod_{i:l_i>0} \partial_{\lambda_i}^{l_i} - \prod_{i:l_i<0} \partial_{\lambda_i}^{-l_i}$$

for $\underline{l} \in \mathbb{L}_A$ and

$$E_k - \beta_k := \sum_{i=1}^{n} a_{ki} \lambda_i \partial_{\lambda_i} - \beta_k$$

for $i = 1, \dots, d$. We call M_A^β a *GKZ−system.*

We are interested to understand when a (localized) hypergeometric $\mathcal{D}-$ module $\mathcal{H} = \frac{\mathbb{C}[t,t^{-1}]\langle \partial_t \rangle}{(P)}$ can be expressed as an inverse image of a $GKZ-$system M_A^β. This is always possible if we allow arbitrary β as has been shown in [CDS17] Corollary 2.8. However, for Hodge theoretic considerations we would like to know more specifically whether β can be chosen to be 0. To be more precise, let $q_1, \dots, q_k \in \mathbb{Z}_{>0}$ with $\gcd(q_1, \dots, q_k) = 1$ and a differential operator

$$P = \gamma \prod_{i=1}^{k} \prod_{j=1}^{q_i-1} (t\partial_t - \frac{j}{q_i}) - t \in \mathbb{C}[t, t^{-1}]\langle \partial_t \rangle$$

be given. Then we have the following result.

Proposition 8.1.7. There is a matrix $A \in \mathrm{Mat}_{(n-1)\times n}(\mathbb{Z})$ and a non-character- istic embedding

$$\iota \colon \mathbb{C}_t^* \to \mathbb{C}^n = \mathrm{Spec}(\mathbb{C}[\lambda_1, \dots, \lambda_n])$$

such that

$$\mathcal{H} := \frac{\mathbb{C}[t, t^{-1}]\langle \partial_t \rangle}{(P)} = \iota^+ \mathrm{M}_A^0.$$

Hence, hypergeometric systems satisfying the exponent condition can be considered as dimensional reductions of GKZ-systems with $\beta = 0$ as parameters.

Proof. By remark 8.1.5 we have that $\mathcal{H} = \mathcal{D}/(P')$ with

$$P' = \gamma \prod_{i=1}^{k} q_i^{-q_i} \prod_{j=0}^{q_i-1} (q_i t\partial_t - j) - t.$$

We set $l := (q_1, \ldots, q_k) \in \mathbb{Z}^k_{>0}$ and choose a matrix $A \in \mathrm{Mat}_{(k-1)\times k}(\mathbb{Z})$, such that $\mathbb{Z}l = \ker(A)$.

Because of $\gcd(q_1, \ldots, q_k) = 1$ there are $(c_1, \ldots, c_k) \in \mathbb{Z}^k$, such that $\sum_{i=1}^{k} c_i q_i = 1$. We define

$$\begin{aligned} \iota\colon \mathbb{C}^* &\to \mathbb{C}^n \\ t &\mapsto (t^{c_1}, \ldots, t^{c_k}). \end{aligned}$$

Notice that it follows from [RS17] Proposition 2.22 that ι is non-characteristic with respect to M_A^0. As $\mathrm{im}(\iota) \subset (\mathbb{C}^*)^k$ we decompose the embedding ι into

$$\mathbb{C}^* \overset{\iota_0}{\hookrightarrow} (\mathbb{C}^*)^k \overset{\iota_1}{\hookrightarrow} \mathbb{C}^k.$$

and compute the inverse image as usual (cf.[HT07] proposition 1.5.11), where $\iota_0(t) = (t^{c_1}, \ldots, t^{c_k})$ and ι_1 is the canonical open embedding. By definition we have (with $\beta = \underline{0} \in \mathbb{Z}^{k-1}$)

$$\mathrm{M}_A^0 = \frac{\mathbb{C}[\lambda_1, \ldots, \lambda_k]\langle \partial_{\lambda_1}, \ldots, \partial_{\lambda_k}\rangle}{(\prod_{i=1}^{k} \partial_{\lambda_i}^{q_i} - 1, E_1, \ldots, E_{k-1})}$$

and by a standard calculation (cf. [HT07] example 1.5.12) we obtain

$$\iota_1^+ M_A^0 = \frac{\mathbb{C}[\lambda_1, \lambda_1^{-1}, \ldots, \lambda_k, \lambda_k^{-1}]\langle \partial_{\lambda_1}, \ldots, \partial_{\lambda_k}\rangle}{(\prod_{i=1}^{k} \partial_{\lambda_i}^{q_i} - 1, E_1, \ldots, E_{k-1})}.$$

As the variables $\lambda_1, \ldots, \lambda_k$ became invertible, we multiply the first generator of the left ideal with $\prod_{i=1}^{k} \lambda_i^{q_i}$ and get

$$\iota_1^+ M_A^0 = \frac{\mathbb{C}[\lambda_1, \lambda_1^{-1}, \ldots, \lambda_k, \lambda_k^{-1}]\langle \partial_{\lambda_1}, \ldots, \partial_{\lambda_k}\rangle}{(\prod_{i=1}^{k} \lambda_i^{q_i}\partial_{\lambda_i}^{q_i} - \prod_{i=1}^{k} \lambda_i^{q_i}, E_1, \ldots, E_{k-1})}.$$

Using the relation $[\partial_{\lambda_i}, \lambda_i] = 1$, we conclude $\lambda_i^{q_i}\partial_{\lambda_i}^{q_i} = \prod_{j=0}^{q_i-1}(\lambda_i\partial_{\lambda_i} - j)$. Hence we get

$$\iota_1^+ M_A^0 = \frac{\mathbb{C}[\lambda_1, \lambda_1^{-1}, \ldots, \lambda_k, \lambda_k^{-1}]\langle \partial_{\lambda_1}, \ldots, \partial_{\lambda_k}\rangle}{(\prod_{i=1}^{k} \prod_{j=0}^{q_i-1}(\lambda_i\partial_{\lambda_i} - j) - \prod_{i=1}^{k} \lambda_i^{q_i}, E_1, \ldots, E_{k-1})}.$$

We claim that

$$\iota_0^+(\iota_1^+ \mathrm{M}_A^0) = \frac{\mathbb{C}[t,t^{-1}]\langle\partial_t\rangle}{(\prod_{i=1}^k \prod_{j=0}^{q_i-1}(q_i t\partial_t - j) - t)}.$$

By construction we have the short exact sequence

$$0 \to \mathbb{Z} \xrightarrow{l^T} \mathbb{Z}^k \xrightarrow{A} \mathbb{Z}^{k-1} \to 0$$

and its dual exact sequence

$$0 \to \mathbb{Z}^{k-1} \xrightarrow{A^T} \mathbb{Z}^k \xrightarrow{l} \mathbb{Z} \to 0.$$

The mapping $i : \mathbb{Z} \to \mathbb{Z}^n$, $1 \mapsto (c_1, \ldots, c_k)$ gives a splitting of the morphism l, and we also chose a splitting $B : \mathbb{Z}^k \to \mathbb{Z}^{k-1}$ such that $B \circ i = 0$. We conclude that (B, l) is inverse to (A^T, i). If we denote for a $m \times n$ matrix with integer values by $\exp(X)$ the monomial map from $(\mathbb{C}^*)^m$ to $(\mathbb{C}^*)^n$ where the exponents are given by the entries of the matrix, then we have

$$\exp(B, l)_+ = \exp(A^T, i)^+,$$

hence $\exp(l)_+ = \exp(i)^+$. Now $\exp(i)$ is nothing but ι_0, and we conclude, by putting $\widetilde{l} := \exp(l) : (\mathbb{C}^*)^k \to \mathbb{C}^*$ that

$$\iota_0^+(\iota_1^+ M_A^0) = \widetilde{l}_+ \iota_1^+ M_A^0.$$

In particular, since ι_0 is non-characteristic with respect to $\iota_1^+ M_A^0$ (as ι is so with respect to M_A^0), we see that $\widetilde{l}_+ \iota_1^+ M_A^0$ is a single degree module. Now it is an easy calculation using the definition of the direct image functor that

$$\iota_0^+(\iota_1^+ M_A^0) = \widetilde{l}_+ \iota_1^+ M_A^0 = \frac{\mathbb{C}[t,t^{-1}]\langle\partial_t\rangle}{(\prod_{i=1}^k \prod_{j=0}^{q_i-1}(q_i t\partial_t - j) - t)}.$$

□

Remark 8.1.8. In the sequel, when considering examples, we will say that the invariant subspace condition holds if the module $G_1(*D)$ is isomorphic to a hypergeometric system satifying this condition. By remark 8.1.3, this amounts to check whether the Bernstein-Sato polynomial $b_h(s)$, *after shifting all of its roots in the intervall* $[0,1)$, is of the form $P = \prod_{i=1}^k \prod_{j=1}^{q_i-1}(t\partial_t - \frac{j}{q_i}) \in \mathbb{C}[t,t^{-1}]\langle\partial_t\rangle$.

8.2 Semi-invariants and spectrum for A_n

This serie was discussed in [GMS09] §6 and we include it for completeness. The A_{n+1}–quiver is

$$1 \xrightarrow{\alpha_1} 2 \xrightarrow{\alpha_2} \dots \xrightarrow{\alpha_{n-1}} n \xrightarrow{\alpha_n} n+1$$

and only the highest root $\mathbf{d} = (1, \dots, 1)$ is scincere (cf. [Bou81] p.248). Obviously $\mathbf{d}$ is a real Schur root with Tit's form equal to one. Let $x_1, \dots, x_n$ denote coordinates on the representation space $\mathrm{Rep}(A_{n+1}, \mathbf{d})$. Then the general representation is given by

$$\mathbb{C}^1 \xrightarrow{A_1} \mathbb{C}^1 \xrightarrow{A_2} \dots \xrightarrow{A_{n-2}} \mathbb{C}^1 \xrightarrow{A_n} \mathbb{C}^1 ,$$

i.e. $A_i = (x_i)$. It is easy to see that the fundamental semi-invariants are $h_i = det(A_i) = x_i$ and the canonical equation is $h = x_1 \cdot \dots \cdot x_n$, i.e. the discriminant of A_{n+1} is the normal crossing divisor.

Proposition 8.2.1. ([GMS09]) The spectrum of A_{n+1} is $(0, \dots, \deg(h)-1)$ and the Sato-Bernstein polynomial of h is $b_h(s) = (s+1)^n$.

Corollary 8.2.2. The A_n–quiver satisfies the exponent condition 8.1.4 and the invariant subspace conjecture 8.1.1.

Proof. In the case of A_{n+1} the group $\mathrm{PGL}(A_{n+1}, \mathbf{d})$ is a torus and the linear space W_T can be chosen to be the representation space $\mathrm{Rep}(A_{n+1}, \mathbf{d})$ itself. Hence the statement $(C1)$ of the invariant subspace conjecture 8.1.1 is satisfied. But then the statements $(C2)$ and $(C3)$ are trivially satisfied, as $h_{|W_T} = h$ is a monomial and the multiplicity of the root -1 of the Bernstein-Sato polynomial $b_h(s)$ is equal to $\dim(\mathrm{PGL}(A_{n+1}, \mathbf{d}))$ by proposition 8.2.1.
Moreover -1 is the only root of $b_h(s)$, hence the exponent condition 8.1.4 is also trivially satisfied. □

Theorem 8.2.3. Let Q be an connected quiver without oriented cycles. If the dimension vector $\mathbf{d} = (1, \dots, 1)$ is a real Schur root and satisfies $q_Q(\mathbf{d}) = 1$, then the canonical equation of the pair $(Q, \mathbf{d})$ is a normal crossing divisor and in particular it is equivalent to the case A_{n+1}.

Proof. In the situation of the theorem it follows directly from the lemma 7.4.3 and Kac's result (7.3.8) that the $\#Q_0 - 1 = \#Q_1$ fundamental semi-invariants are given by

$$h_i = \det(A_\alpha),$$

where A_α denotes the 1×1 matrix, whose entry is the α−th coordinate function of $\mathrm{Rep}(Q, \mathbf{d})$. Moreover because the canonical equation has degree equal to $\dim(\mathrm{Rep}(Q, \mathbf{d}))$, which is in this situation obviously equal to $\#Q_1$, it follows that the canonical equation is the product of the coordinate functions (all with multiplicity 1). □

8.3 Semi-invariants and spectrum for D_n

We consider the Dynkin quiver D_n ($n \geq 4$) with the following ordering of nodes and orientation of arrows:

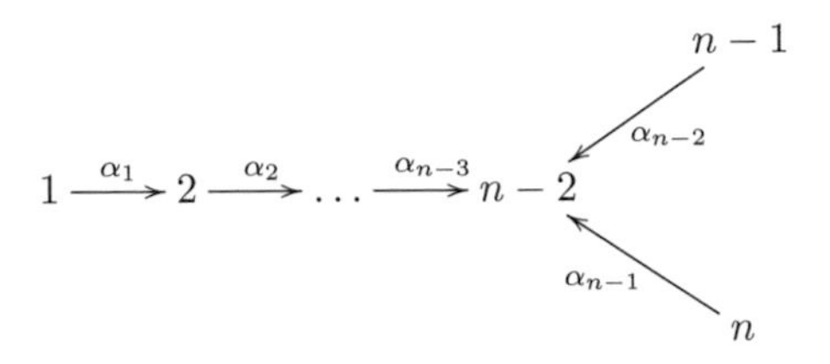

By [Bou81] p. 256 the highest root is $\mathbf{d} = (1, 2, \dots, 2, 1, 1)$. We denote by h the canonical equation of $(D_n, \mathbf{d})$.

Proposition 8.3.1. ([GMS09] section 6) The Bernstein-Sato polynomial of h is

$$b_h(s) = (s + \frac{2}{3})^{n-3}(s+1)^{2n-4}(s + \frac{4}{3})^{n-3}$$

and the spectrum (at $t = 0$ and $t \neq 0$) is

$\frac{4n+3i-13}{3}$	$i-1$	$\frac{3i-4n-13}{3}$
$i = 1, \dots, n-3$	$i = n-2, \dots, 3n-7$	$i = 3n-6, \dots, 4n-10$

Corollary 8.3.2. The pair $(D_n, \mathbf{d})$ satisfies the exponent condition 8.1.4 and the invariant subspace conjecture 8.1.1.

Proof. The exponent condition 8.1.4 is obviously satisfied by proposition 8.3.1, as after shifting the roots of the Bernstein-Sato polynomial into the interval $[0, 1)$, the shifted roots $\frac{1}{3}$ and $\frac{2}{3}$ both appear with multiplicity $n-3$ and the root 0 with multiplicity $2n-4$.
For the invariant subspace conjecture 8.1.1 we introduce some additional

notation. We denote by

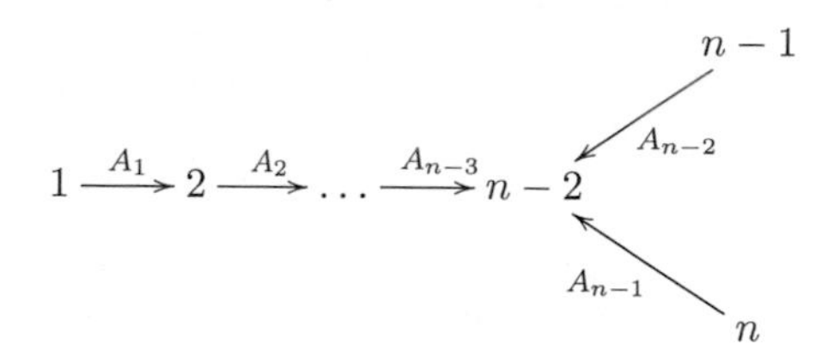

the general representation of $\mathrm{Rep}(D_n, \mathbf{d})$, which we identify with $\mathrm{Mat}_{2\times 3+2(n-4)}$ by

$$(A_1, \dots, A_{n-1}) \mapsto (A_1 | \dots | A_{n-1}).$$

Let $\tilde{T} \leq \mathrm{GL}(D_n, \mathbf{d})$ be the maximal torus consisting of diagonal matrices and let $T = \mathrm{im}(\tilde{T})$, where $\pi\colon \mathrm{GL}(D_n, \mathbf{d}) \to \mathrm{PGL}(D_n, \mathbf{d})$ is the canonical projection.
It follows from a direct calculation that the isotropy group $\mathrm{GL}(D_n, \mathbf{d})_M$ of the representation

$$M = (\begin{pmatrix} 1 \\ 0 \end{pmatrix}, \begin{pmatrix} 1 & 0 \\ 0 & 1 \end{pmatrix}, \begin{pmatrix} 1 & 0 \\ 0 & 1 \end{pmatrix}, \dots, \begin{pmatrix} 1 & 0 \\ 0 & 1 \end{pmatrix}, \begin{pmatrix} 0 \\ 1 \end{pmatrix}, \begin{pmatrix} 1 \\ 1 \end{pmatrix})$$

is isomorphic to $\mathbb{C}^*$ and hence by remark 7.2.9 the $\mathrm{PGL}(D_n, \mathbf{d})-$orbit of M is open in $\mathrm{Rep}(D_n, \mathbf{d})$. Furthermore it follows from a direct calculation that $W_T = \overline{T.M}$, the closure of the $T-$orbit, is given by the matrices

$$\begin{pmatrix} * & * & 0 & * & 0 & \dots & * & 0 & 0 & * \\ 0 & 0 & * & 0 & * & \dots & 0 & * & * & * \end{pmatrix} \in \mathrm{Mat}_{2\times 3+2(n-4)},$$

where the $*$s are arbitrary complex numbers. In particular W_T is a vector space. Because M is a brick, we have $W_T \not\subset D$. Indeed the divisor D consists of decomposable representations, but the representation M is indecomposable. We conclude that the statement $(C1)$ of the invariant subspace conjecture is satisfied.
Let us prove the statement $(C2)$ next. First we note that if the restrictions of fundamental semi-invariants $(h_i)_{|W_T}$ to W_T are monomials, then the restriction of any defining equation $(h)_{|W_T}$ of the discriminant is a monomial. Indeed as h is a semi-invariant, it is a product of the fundamental semi-invariants by theorem 1.1.10.
We conclude from the discussion in section 7.4 that the $n-1$ fundamental semi-invariants are

$$\begin{array}{rcll} h_i &=& \det(A_{i+1}) & \text{, for } i = 1, \dots, n-4 \\ h_{n-3} &=& \det(A_{n-3} \cdot \ldots \cdot A_1 | A_{n-2}) & \\ h_{n-2} &=& \det(A_{n-3} \cdot \ldots \cdot A_1 | A_{n-1}) & \\ h_{n-1} &=& \det(A_{n-2} | A_{n-1}) & \end{array} .$$

Hence if we write $A_l = (a_{ij}^{(l)})$, where $a_{ij}^{(l)}$ are the coordinate functions on $\operatorname{Rep}(D_n, \mathbf{d})$, the restrictions of the fundamental semi-invariants to W_T are

$$\begin{aligned} (h_i)_{|W_T} &= a_{11}^{(i+1)} a_{22}^{(i+2)}, \text{ for } i = 1, \dots, n-4 \\ (h_{n-3})_{|W_T} &= \det\left(\begin{pmatrix} \prod_{i=2}^{n-3} a_{11}^{(i)} & 0 \\ 0 & \prod_{i=2}^{n-3} a_{22}^{(i)} \end{pmatrix} \begin{pmatrix} a_{11}^{(1)} \\ 0 \end{pmatrix} \middle| \begin{pmatrix} 0 \\ a_{21}^{(n-2)} \end{pmatrix} \right) \\ &= a_{11}^{(1)} a_{21}^{(n-2)} \prod_{i=2}^{n-3} a_{11}^{(i)} \\ (h_{n-2})_{|W_T} &= \det\left(\begin{pmatrix} \prod_{i=2}^{n-3} a_{11}^{(i)} & 0 \\ 0 & \prod_{i=2}^{n-3} a_{22}^{(i)} \end{pmatrix} \begin{pmatrix} a_{11}^{(1)} \\ 0 \end{pmatrix} \middle| \begin{pmatrix} a_{11}^{(n-1)} \\ a_{21}^{(n-1)} \end{pmatrix} \right) \\ &= a_{11}^{(1)} a_{21}^{(n-1)} \prod_{i=2}^{n-3} a_{11}^{(i)} \\ (h_{n-1})_{|W_T} &= \det\left(\begin{pmatrix} 0 & a_{11}^{(n-1)} \\ a_{21}^{(n-2)} & a_{21}^{(n-1)} \end{pmatrix} \right) = -a_{21}^{(n-2)} a_{11}^{(n-1)} \end{aligned}$$

We see that the restrictions of the fundamental semi-invariants are monomial and hence the restriction $(h)_{|W_T}$ is a monomial.
More precisely, by comparing the degrees we conclude that $h = \prod_{i=1}^{n-1} h_i$ is the cannonical equation[2].
At last as $\dim(T) = 2n-4$ the statement $(C3)$ of the invariant subspace conjecture 8.1.1 is satisfied by proposition 8.3.1.
This finishes the proof of the invariant subspace conjecture 8.1.1. □

Remark 8.3.3. We remark that by [BM06] proposition 7.8 changing the orientation of D_n leads to an isomorphic divisor. In particular the statements in this subsections hold for any choice of orientation for the Dynkin quiver D_n.

[2] This would also follow from [BM06] corollary 5.5 and proposition 6.3.

8.4 Semi-invariants and spectrum for E_6

In this subsection we consider the Dynkin quiver E_6 with the following ordering of nodes:

$$\begin{array}{ccccccccc} 1 & \xrightarrow{\alpha} & 2 & \xrightarrow{\beta} & 3 & \xrightarrow{\gamma} & 4 & \xrightarrow{\delta} & 5 \\ & & & & \uparrow{\scriptstyle \epsilon} & & & & \\ & & & & 6 & & & & \end{array}.$$

By [Bou81] p.259 ff. the positive sincere roots are

$\mathbf{d}^1: \begin{matrix} 1 & 1 & 2 & 1 & 1 \\ & & 1 & & \end{matrix}$	$\mathbf{d}^2: \begin{matrix} 1 & 2 & 2 & 1 & 1 \\ & & 1 & & \end{matrix}$
$\mathbf{d}^3: \begin{matrix} 1 & 1 & 2 & 2 & 1 \\ & & 1 & & \end{matrix}$	$\mathbf{d}^4: \begin{matrix} 1 & 2 & 2 & 2 & 1 \\ & & 1 & & \end{matrix}$
$\mathbf{d}^5: \begin{matrix} 1 & 2 & 3 & 2 & 1 \\ & & 1 & & \end{matrix}$	$\mathbf{d}^6: \begin{matrix} 1 & 2 & 3 & 2 & 1 \\ & & 2 & & \end{matrix}$

Because the Tits form of each of these roots is equal to one, we have:

Proposition 8.4.1. Every root $\mathbf{d}^i$, $i = 1, \ldots, 6$, from above is a real Schur root and each pair $(\mathrm{PGL}(E_6, \mathbf{d}^i), \mathrm{Rep}(E_6, \mathbf{d}^i))$ is a regular reductive prehomogeneous vector space.

Proof. One easily verifies that we have for $i = 1, \ldots, 6$

$$q_{E_6}(\mathbf{d}^i) = 1.$$

Hence by theorem 7.2.15. each $\mathbf{d}^i$ is a real Schur root and by corollary 7.2.16 each pair $(\mathrm{PGL}(E_6, \mathbf{d}^i), \mathrm{Rep}(E_6, \mathbf{d}^i))$ is a reductive prehomogeneous vector space.
Moreover by theorem 7.2.20 the complement of the open orbit is a divisor, which is by theorem 1.3.2 equivalent to the regularity of the reductive prehomogeneous vector space $(\mathrm{PGL}(E_6, \mathbf{d}^i), \mathrm{Rep}(E_6, \mathbf{d}^i))$. □

Remark 8.4.2. We remark that the non sincere positive roots of E_6 can be considered as sincere roots of the Dynkin quivers D_4 and D_5.

Remark 8.4.3. By the result of V.Kac (7.3.8) there are precisely $5 = (E_6)_0 - 1$ fundamental semi-invariants for each pair $(E_6, \mathbf{d}^i)$, $i = 1, \ldots, 6$.

Remark 8.4.4. We briefly recall that we find the (fundamental) semi-invariants by Schofield's Method as follows: First we find orthogonal roots $\mathbf{e}$, i.e. $\mathbf{e} \in \mathbb{N}^{(E_6)_0}$ with

$$\langle \mathbf{e}, \mathbf{d}^i \rangle_{E_6} = \mathbf{e} E \mathbf{d}^T = 0$$

where

$$E = \begin{pmatrix} 1 & -1 & 0 & 0 & 0 & 0 \\ 0 & 1 & -1 & 0 & 0 & 0 \\ 0 & 0 & 1 & -1 & 0 & 0 \\ 0 & 0 & 0 & 1 & -1 & 0 \\ 0 & 0 & 0 & 0 & 1 & 0 \\ 0 & 0 & -1 & 0 & 0 & 1 \end{pmatrix}$$

is Euler matrix for this orientation. Then for a generic representation $N \in \mathrm{Rep}(E_6, \mathbf{e})$ and the general representation[3] M of $\mathrm{Rep}(E_6, \mathbf{d}^i)$ the determinant of the map

$$c_{M,N} = \begin{pmatrix} S_2 a - AS_1 \\ S_3 b - BS_2 \\ S_4 c - CS_3 \\ S_5 d - DS_4 \\ S_3 e - ES_6 \end{pmatrix}$$

is a semi-invariant. That $c_{M,N}$ is as above, follows directly form the diagram

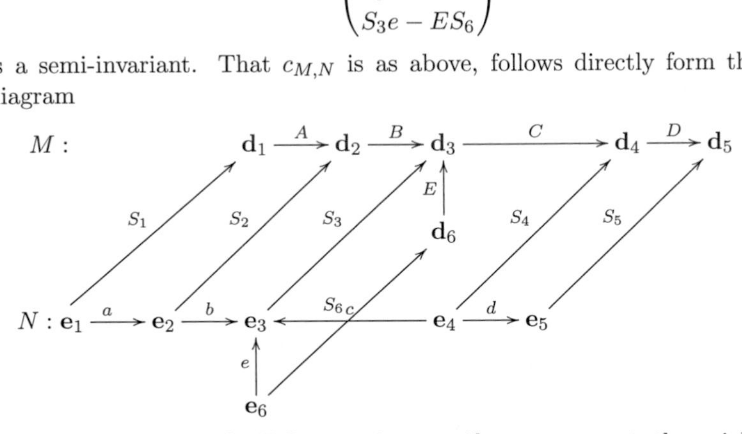

Moreover we can use the Euler matrix to verify or to compute the weight of the semi-invariant c^N using lemma 7.3.2.

In the following we give for each root $\mathbf{d}^i$ the set of fundamental semi-invariants, together with their degree and weights, and the corresponding orthogonal root $\mathbf{e}$. We give the Bernstein-Sato polynomial of the canonical equation Δ and the Spectrum at $t = 0$ and $t \neq 0$, which is computed by using the algorithms explained in the introduction of this chapter. We end each example by giving a representation, which is the key ingredient to solve the invariant subspace conjecture 8.1.1 in this particular example. Moreover if the degree of Δ is not to big, we also give the equation of the restriction $\Delta_{|W_T}$.

[3] I.e. the entries of the matrices in M are the coordinate functions of $\mathrm{Rep}(E_6, \mathbf{d}^i)$.

The root $\mathbf{d} = \mathbf{d}^1$: $\begin{matrix} \mathbf{1} & \mathbf{1} & \mathbf{2} & \mathbf{1} & \mathbf{1} \\ & & \mathbf{1} & & \end{matrix}$

We have $\dim(\mathrm{Rep}(E_6, \mathbf{d}^1)) = 8$ and the relation for a dimension vector $\mathbf{e}$ to be orthogonal to $\mathbf{d}^1$ is $0 = \langle \mathbf{e}, \mathbf{d}^1 \rangle_{E_6} = -e_2 + e_3 + e_5 - e_6$.

Semi-invariant	Degree	$\mathrm{Root}^{\perp \mathbf{d}}$	Weight
$h_1 = \det(A)$	1	$\begin{matrix} 1 & 0 & 0 & 0 & 0 \\ & & 0 & & \end{matrix}$	$\begin{matrix} 1 & -1 & 0 & 0 & 0 \\ & & 0 & & \end{matrix}$
$h_2 = \det(D)$	1	$\begin{matrix} 0 & 0 & 0 & 1 & 0 \\ & & 0 & & \end{matrix}$	$\begin{matrix} 0 & 0 & 0 & 1 & -1 \\ & & 0 & & \end{matrix}$
$h_3 = \det(CE)$	2	$\begin{matrix} 0 & 0 & 1 & 0 & 0 \\ & & 1 & & \end{matrix}$	$\begin{matrix} 0 & 0 & 0 & -1 & 0 \\ & & 1 & & \end{matrix}$
$h_4 = \det(CB)$	2	$\begin{matrix} 0 & 1 & 1 & 0 & 0 \\ & & 0 & & \end{matrix}$	$\begin{matrix} 0 & 1 & 0 & -1 & 0 \\ & & 0 & & \end{matrix}$
$h_5 = \det(B\|E)$	2	$\begin{matrix} 0 & 1 & 1 & 1 & 1 \\ & & 1 & & \end{matrix}$	$\begin{matrix} 0 & 1 & -1 & 0 & 0 \\ & & 1 & & \end{matrix}$
$\Delta = h_1 \dots h_5$	8		$\begin{matrix} 1 & 1 & -1 & -1 & -1 \\ & & 2 & & \end{matrix}$

The Bernstein-Sato polynomial is

$$b_\Delta(s) = (s + \frac{2}{3})^1 (s+1)^6 (s + \frac{4}{3})^1.$$

Spectrum (at $t = 0$ and $t \neq 0$) is

$$(2\frac{2}{3}, 1, 2, 3, 4, 5, 6, 4\frac{1}{3})$$

For the invariant subspace conjecture 8.1.1 we consider the representation

$$M : \mathbb{C}^1 \xrightarrow{(1)} \mathbb{C}^1 \xrightarrow{\begin{pmatrix}1\\0\end{pmatrix}} \mathbb{C}^2 \xrightarrow{(1\ \ 1)} \mathbb{C}^1 \xrightarrow{(1)} \mathbb{C}^1, \qquad \mathbb{C}^1 \xrightarrow{\begin{pmatrix}0\\1\end{pmatrix}} \mathbb{C}^2$$

and the maximal torus T given by the diagonal matrices of $\mathrm{GL}(E_6, \mathbf{d}^1)$, i.e.

$$T = \{(\lambda_1, \lambda_2, \begin{pmatrix} \lambda_3 & 0 \\ 0 & \lambda_4 \end{pmatrix}, \lambda_5, \lambda_6, \lambda_7) \mid \lambda_i \in \mathbb{C}^*\}.$$

We denote by $A_1, \dots, A_7$ be the standard basis of $\mathfrak{t} = \mathrm{Lie}(T)$, i.e.

$$A_1 = (1,0,\begin{pmatrix}0&0\\0&0\end{pmatrix},0,0,0),\quad A_2 = (0,1,\begin{pmatrix}0&0\\0&0\end{pmatrix},0,0,0),$$
$$A_3 = (0,0,\begin{pmatrix}1&0\\0&0\end{pmatrix},0,0,0),\quad A_4 = (0,0,\begin{pmatrix}0&0\\0&1\end{pmatrix},0,0,0),$$
$$A_5 = (0,0,\begin{pmatrix}0&0\\0&0\end{pmatrix},1,0,0),\quad A_6 = (0,0,\begin{pmatrix}0&0\\0&0\end{pmatrix},0,1,0),$$
$$A_7 = (0,0,\begin{pmatrix}0&0\\0&0\end{pmatrix},0,0,1).$$

and obtain a set of generators of W_T by $M_i = A_i.M$, i.e.

$$M_1 = (-1,\begin{pmatrix}0\\0\end{pmatrix},\begin{pmatrix}0&0\end{pmatrix},0,\begin{pmatrix}0\\0\end{pmatrix}),\quad M_2 = (1,-\begin{pmatrix}1\\0\end{pmatrix},\begin{pmatrix}0&0\end{pmatrix},0,\begin{pmatrix}0\\0\end{pmatrix}),$$
$$M_3 = (0,\begin{pmatrix}1\\0\end{pmatrix},-\begin{pmatrix}1&0\end{pmatrix},0,\begin{pmatrix}0\\0\end{pmatrix}),\quad M_4 = (0,\begin{pmatrix}0\\0\end{pmatrix},-\begin{pmatrix}0&1\end{pmatrix},0,\begin{pmatrix}0\\1\end{pmatrix}),$$
$$M_5 = (0,\begin{pmatrix}0\\0\end{pmatrix},\begin{pmatrix}1&1\end{pmatrix},-1,\begin{pmatrix}0\\0\end{pmatrix}),\quad M_6 = (0,\begin{pmatrix}0\\0\end{pmatrix},\begin{pmatrix}0&0\end{pmatrix},1,\begin{pmatrix}0\\0\end{pmatrix}),$$
$$M_7 = (0,\begin{pmatrix}0\\0\end{pmatrix},\begin{pmatrix}0&0\end{pmatrix},0,-\begin{pmatrix}0\\1\end{pmatrix}).$$

We write an element of W_T as a vector, where the components are ordered as

$$(1,\begin{pmatrix}2\\3\end{pmatrix},\begin{pmatrix}4&5\end{pmatrix},6,-\begin{pmatrix}7\\8\end{pmatrix}).$$

We easily see that the elements

$$\begin{aligned}
a_{11} &= (1,0,0,0,0,0,0,0)\\
b_{11} &= (0,1,0,0,0,0,0,0)\\
c_{11} &= (0,0,0,1,0,0,0,0)\\
c_{12} &= (0,0,0,0,1,0,0,0)\\
d_{11} &= (0,0,0,0,0,1,0,0)\\
e_{21} &= (0,0,0,0,0,0,0,1)
\end{aligned}$$

form a basis of W_T and for simplicity we denote the coordinate functions by the same symbols. We conclude from a direct calculation that the restriction of Δ to W_T is

$$\Delta_{|W_T} = a_{11}b_{11}^2c_{11}c_{12}d_{11}e_{21}^2.$$

The root $\mathbf{d} = \mathbf{d}^2$: $\begin{matrix} \mathbf{1} & \mathbf{2} & \mathbf{2} & \mathbf{1} & \mathbf{1} \\ & & \mathbf{1} & & \end{matrix}$

We have $\dim(\mathrm{Rep}(E_6, \mathbf{d}^2)) = 11$ and the relation for a dimension vector $\mathbf{e}$ to be orthogonal to $\mathbf{d}^2$ is $0 = -e_1 + e_3 + e_5 - e_6$.

Semi-Invariant	Degree	Root$^{\perp \mathbf{d}}$	Weight
$h_1 = \det(D)$	1	0 0 0 1 0 0	0 0 0 1 -1 0
$h_2 = \det(B)$	2	0 1 0 0 0 0	0 1 -1 0 0 0
$h_3 = \det(CE)$	2	0 0 1 0 0 1	0 0 0 -1 0 1
$h_4 = \det(CBA)$	3	1 1 1 0 0 0	1 0 0 -1 0 0
$h_5 = \det(BA\|E)$	3	1 1 1 1 1 1	1 0 -1 0 0 1
$\Delta = h_1 \dots h_5$	11		2 1 -2 -1 -1 2

The Bernstein-Sato polynomial is

$$b_\Delta(s) = (s + \frac{2}{3})^2 (s+1)^7 (s + \frac{4}{3})^2.$$

The Spectrum (at $t = 0$ and $t \neq 0$) is

$$(\frac{11}{3}, \frac{14}{3}, 2, \dots, 8, \frac{16}{3}, \frac{19}{3})$$

If we consider for the invariant subspace conjecture 8.1.1 the representation

$$M : \begin{array}{ccccccccc} \mathbb{C}^1 & \xrightarrow{\begin{pmatrix}1\\0\end{pmatrix}} & \mathbb{C}^2 & \xrightarrow{\begin{pmatrix}1 & 0\\0 & 1\end{pmatrix}} & \mathbb{C}^2 & \xrightarrow{\begin{pmatrix}1 & 1\end{pmatrix}} & \mathbb{C}^1 & \xrightarrow{1} & \mathbb{C}^1 , \\ & & & & \uparrow{\scriptstyle \begin{pmatrix}0\\1\end{pmatrix}} & & & & \\ & & & & \mathbb{C}^1 & & & & \end{array}$$

then the restriction of Δ to the subspace $W_T := \overline{T.M}$ is

$$\Delta_{|W_T} = a_{11}^2 b_{11}^3 b_{22} c_{11} c_{12} d_1 e_{21}^2.$$

The root $\mathbf{d} = \mathbf{d}^3$: $\begin{smallmatrix} \mathbf{1} & \mathbf{1} & \mathbf{2} & \mathbf{2} & \mathbf{1} \\ & & \mathbf{1} & & \end{smallmatrix}$

We have $\dim(\text{Rep}(E_6, \mathbf{d}^3)) = 11$ and the relation for a dimension vector $\mathbf{e}$ to be orthogonal to $\mathbf{d}^3$ is $0 = -e_2 + e_4 + e_5 - e_6$.

Semi-invariant	Degree	Root$^{\perp \mathbf{d}}$	Weight
$h_1 = \det(A)$	1	1 0 0 0 0 0	1 -1 0 0 0 0
$h_2 = \det(C)$	2	0 0 1 0 0 0	0 0 1 -1 0 0
$h_3 = \det(B\|E)$	2	0 1 1 1 1 1	0 1 -1 0 0 1
$h_4 = \det(DCE)$	3	0 0 1 1 0 1	0 0 0 0 -1 1
$h_5 = \det(DCB)$	3	0 1 1 1 0 0	0 1 0 0 -1 0
$\Delta = h_1 \dots h_5$	11		1 1 0 -1 -2 2

Remark 8.4.5. This divisor is isomorphic to the one defined by the real Schur root $\mathbf{d}^2$ and for completeness we note:
The Bernstein-Sato polynomial is

$$b_\Delta(s) = (s + \frac{2}{3})^2 (s+1)^7 (s + \frac{4}{3})^2.$$

The Spectrum (at $t = 0$ and $t \neq 0$) is

$$(\frac{11}{3}, \frac{14}{3}, 2, \dots, 8, \frac{16}{3}, \frac{19}{3}).$$

If we consider for the invariant subspace conjecture 8.1.1 the representation

$$M : \begin{array}{ccccccccc} \mathbb{C}^1 & \xrightarrow{1} & \mathbb{C}^1 & \xrightarrow{\begin{pmatrix}1\\0\end{pmatrix}} & \mathbb{C}^2 & \xrightarrow{\begin{pmatrix}1&0\\0&1\end{pmatrix}} & \mathbb{C}^2 & \xrightarrow{\begin{pmatrix}1&1\end{pmatrix}} & \mathbb{C}^1 \\ & & & & \uparrow{\scriptstyle\begin{pmatrix}0\\1\end{pmatrix}} & & & & \\ & & & & \mathbb{C}^1 & & & & \end{array},$$

then the restriction of Δ to the subspace $W_T := \overline{T.M}$ is

$$\Delta =_{|W_T} = a_{11} b_{11}^2 c_{11}^2 c_{22}^2 d_{11} d_{12} e_{21}^2.$$

The root $\mathbf{d} = \mathbf{d}^4$: $\begin{matrix} \mathbf{1} & \mathbf{2} & \mathbf{2} & \mathbf{2} & \mathbf{1} \\ & & \mathbf{1} & & \end{matrix}$

We have $\dim(\mathrm{Rep}(E_6, \mathbf{d}^4)) = 14$ and the relation for a dimension vector $\mathbf{e}$ to be orthogonal to $\mathbf{d}^4$ is $0 = -e_1 + e_4 + e_5 - e_6$.

Semi-invariant	Degree	$\mathrm{Root}^{\perp \mathbf{d}}$	Weight
$h_1 = \det(B)$	2	0 1 0 0 0 0	0 1 -1 0 0 0
$h_2 = \det(C)$	2	0 0 1 0 0 0	0 0 1 -1 0 0
$h_3 = \det(DCE)$	3	0 0 1 1 0 1	0 0 0 0 -1 1
$h_4 = \det(BA\|E)$	3	1 1 1 1 1 1	1 0 -1 0 0 1
$h_5 = \det(DCBA)$	4	1 1 1 1 0 0	1 0 0 0 -1 0
$\Delta = h_1 \dots h_5$	14		2 1 -1 -1 -2 2

The Bernstein-Sato polynomial is

$$b_\Delta(s) = (s + \frac{2}{3})^3 (s+1)^8 (s + \frac{4}{3})^3.$$

The Spectrum (at $t = 0$ and $t \neq 0$) is

$$(\frac{14}{3}, \frac{17}{3}, \frac{20}{3}, 3, \dots, 10, \frac{19}{3}, \frac{22}{3}, \frac{25}{3})$$

If we consider for the invariant subspace conjecture 8.1.1 the representation

$$M: \ \mathbb{C}^1 \xrightarrow{\begin{pmatrix}1\\0\end{pmatrix}} \mathbb{C}^1 \xrightarrow{\begin{pmatrix}1&0\\0&1\end{pmatrix}} \mathbb{C}^2 \xrightarrow{\begin{pmatrix}1&0\\0&1\end{pmatrix}} \mathbb{C}^2 \xrightarrow{\begin{pmatrix}1&1\end{pmatrix}} \mathbb{C}^1, \qquad \mathbb{C}^1 \xrightarrow{\begin{pmatrix}0\\1\end{pmatrix}} \mathbb{C}^2 \ (\text{vertical arrow into the middle } \mathbb{C}^2)$$

then the restriction of Δ to the subspace $W_T := \overline{T.M}$ is

$$\Delta_{|W_T} = a_{11}^2 b_{11}^3 b_{22} c_{11} c_{22}^2 d_{11} d_{12} e_{21}^2.$$

The root $\mathbf{d} = \mathbf{d}^5$: $\begin{matrix} \mathbf{1} & \mathbf{2} & \mathbf{3} & \mathbf{2} & \mathbf{1} \\ & & \mathbf{1} & & \end{matrix}$

We have $\dim(\mathrm{Rep}(E_6, \mathbf{d}^5)) = 19$ and the relation for a dimension vector $\mathbf{e}$ to be orthogonal to $\mathbf{d}^5$ is $0 = -e_1 - e_2 + e_3 + e_4 + e_5 - 2e_6$.

Semi-invariant	Degree	Root$^{\perp \mathbf{d}}$	Weight
$h_1 = \det(DCE)$	3	0 0 1 1 0 1	0 0 0 0 -1 1
$h_2 = (B\|E)$	3	0 1 1 1 1 1	0 1 -1 0 0 1
$h_3 = \det(CB)$	4	0 1 1 0 0 0	0 1 0 -1 0 0
$h_4 = \det(DCBA)$	4	1 1 1 1 0 0	1 0 0 0 -1 0
$h_5 = \det(CBA\|CE)$	5	1 1 2 1 1 1	1 0 0 -1 0 1
$\Delta = h_1 \ldots h_5$	19		2 2 -1 -2 -2 3

The Bernstein-Sato polynomial is

$$b_\Delta(s) = (s + \frac{3}{5})(s + (\frac{2}{3})^3)(s + \frac{4}{5})(s + 1)^9(s + \frac{6}{5})(s + \frac{4}{3})^3(s + \frac{7}{5}).$$

The Spectrum (at $t = 0$ and $t \neq 0$) is

$$(\frac{38}{5}, \frac{22}{3}, \frac{25}{3}, \frac{28}{3}, \frac{39}{5}, 5, \ldots, 13, \frac{51}{5}, \frac{26}{3}, \frac{29}{3}, \frac{32}{3}, \frac{52}{5}).$$

If we consider for the invariant subspace conjecture 8.1.1 the representation

$$M : \mathbb{C}^1 \xrightarrow{\begin{pmatrix}1\\0\end{pmatrix}} \mathbb{C}^2 \xrightarrow{\begin{pmatrix}1&0\\0&1\\1&1\end{pmatrix}} \mathbb{C}^3 \xrightarrow{\begin{pmatrix}1&0&1\\0&1&1\end{pmatrix}} \mathbb{C}^2 \xrightarrow{\begin{pmatrix}0&1\end{pmatrix}} \mathbb{C}^1 .$$
$$\mathbb{C}^1 \xrightarrow{\begin{pmatrix}1&1&1\end{pmatrix}^T} \mathbb{C}^3 \text{ (vertical arrow into the middle } \mathbb{C}^3)$$

then the restrictions of Δ to the subspace $W_T := \overline{T.M}$ is

$$\Delta_{|W_T} = \overline{a}_{11}^2 \overline{b}_{11}^4 \overline{b}_{22}^2 \overline{c}_{11}^3 \overline{c}_{22}^3 \overline{d}_{12}^2 \overline{e}_{11}^3$$

(seen in $\mathbb{C}[\mathrm{Rep}(E_6, \mathbf{d}_5]/I$, where I is the ideal of $W_T = \overline{T.M}$).

The root $\mathbf{d} = \mathbf{d}^6$: $\begin{matrix} \mathbf{1} & \mathbf{2} & \mathbf{3} & \mathbf{2} & \mathbf{1} \\ & & \mathbf{2} & & \end{matrix}$

We have $\dim(\mathrm{Rep}(E_6, \mathbf{d}^6)) = 22$ and a dimension vector $\mathbf{e}$ is orthogonal to $\mathbf{d}^6$, if it satisfies the relation $0 = -e_1 - e_2 + e_3 + e_4 + e_5 - e_6$.

Semi-invariant	Degree	Root$^{\perp \mathbf{d}}$	Weight
$h_1 = \det(DCBA)$	4	$\begin{matrix} 1 & 1 & 1 & 1 & 0 \\ & & 0 & & \end{matrix}$	$\begin{matrix} 1 & 0 & 0 & 0 & -1 \\ & & 0 & & \end{matrix}$
$h_2 = \det(CB)$	4	$\begin{matrix} 0 & 1 & 1 & 0 & 0 \\ & & 0 & & \end{matrix}$	$\begin{matrix} 0 & 1 & 0 & -1 & 0 \\ & & 0 & & \end{matrix}$
$h_3 = \det(CE)$	4	$\begin{matrix} 0 & 0 & 1 & 0 & 0 \\ & & 1 & & \end{matrix}$	$\begin{matrix} 0 & 0 & 0 & -1 & 0 \\ & & 1 & & \end{matrix}$
$h_4 = \det(BA\vert E)$	4	$\begin{matrix} 1 & 1 & 1 & 1 & 1 \\ & & 1 & & \end{matrix}$	$\begin{matrix} 1 & 0 & -1 & 0 & 0 \\ & & 1 & & \end{matrix}$
$h_5 = \det(\Gamma)$	6	$\begin{matrix} 0 & 1 & 1 & 1 & 0 \\ & & 1 & & \end{matrix}$	$\begin{matrix} 0 & 1 & -1 & 0 & -1 \\ & & 1 & & \end{matrix}$
$\Delta = h_1 \dots h_5$	22		$\begin{matrix} 2 & 2 & -2 & -2 & -2 \\ & & 3 & & \end{matrix}$

$$\text{where } \Gamma = \begin{pmatrix} B & \mathbf{1}_3 & 0 & 0 \\ 0 & C & \mathbf{1}_2 & 0 \\ 0 & 0 & D & 0 \\ 0 & \mathbf{1}_3 & 0 & E \end{pmatrix}.$$

The Bernstein-Sato polynomial is

$$b_\Delta(s) = (s + \frac{3}{5})(s + (\frac{2}{3})^4(s + \frac{4}{5})(s+1)^{10}(s + \frac{6}{5})(s + \frac{4}{3})^4(s + \frac{7}{5}).$$

The Spectrum (at $t = 0$ and $t \neq 0$) is

$$(\frac{44}{5}, \frac{25}{3}, \frac{28}{3}, \frac{31}{3}, \frac{34}{3}, \frac{47}{5}, 6, \dots, 15, \frac{58}{5}, \frac{29}{3}, \frac{32}{3}, \frac{35}{3}, \frac{38}{3}, \frac{61}{5})$$

For the invariant subspace conjecture 8.1.1 we consider the representation

$$M : \begin{matrix} \mathbb{C}^1 & \xrightarrow{A} & \mathbb{C}^2 & \xrightarrow{B} & \mathbb{C}^3 & \xrightarrow{C} & \mathbb{C}^2 & \xrightarrow{D} & \mathbb{C}^1, \\ & & & & \uparrow{\scriptstyle E} & & & & \\ & & & & \mathbb{C}^2 & & & & \end{matrix}$$

where the matrices are

$$A = \begin{pmatrix} 1 \\ 0 \end{pmatrix}, \quad B = \begin{pmatrix} 1 & 0 \\ 0 & 1 \\ 1 & 1 \end{pmatrix} \quad C = \begin{pmatrix} 1 & 0 & 1 \\ 0 & 1 & 1 \end{pmatrix}$$

$$D = \begin{pmatrix} 0 & 1 \end{pmatrix} \quad E = \begin{pmatrix} 1 & 0 \\ 1 & 1 \\ 0 & 0 \end{pmatrix} .$$

The restriction of Δ to the space $W_T := \overline{T.M}$ is a monomial of degree 22 in 10 variables.

Evaluation of E_6

We see by inspection that the spectrum of E_6 consists of positive rational numbers, which fulfil the symmetry $\nu_i + \nu_{n+1-i} = \deg(\Delta) - 1$.
This is not a surprise, as by corollary 6.2.10 the spectral numbers correspond to the roots of the Bernstein-Sato polynomial and the roots of the Bernstein-Sato polynomial are symmetric around -1 by [GS10] theorem 1.4.
Until root $\mathbf{d}^5$ we see that only the roots $-\frac{4}{3}, -1, -\frac{2}{3}$ appear and their multiplicity rises as the degree of Δ is rising.
With root $\mathbf{d}^5$ also the roots $-\frac{7}{5}, -\frac{6}{5}, -\frac{4}{5}, -\frac{3}{5}$ appears with multiplicity one. Also the multiplicity of -1 rises by one, but the multiplicity of the roots $-\frac{4}{3}, -\frac{2}{3}$ remains three.
For highest root $\mathbf{d}^6$ the multiplicity of the roots $-\frac{7}{5}, -\frac{6}{5}, -\frac{4}{5}, -\frac{3}{5}$ remains one, while the multiplicity of the others is rising by one.

Remark 8.4.6. We note that the difference of the dimension the representation spaces for $\mathbf{d}^i$ and $\mathbf{d}^{i+1}$, for $i = 1, \ldots, 5$, is in a certain sense equally distributed to the multiplicity of the roots of the Bernstein-Sato polynomial.
Precisely; The difference of $\dim(\operatorname{Rep}(E_6, \mathbf{d}^4)$ and $\operatorname{Rep}(E_6, \mathbf{d}^5)$ is five and in this case the multiplicity of $-\frac{7}{5}, -\frac{6}{5}, -\frac{4}{5}, -\frac{3}{5}$ rises from 0 to 1 and the multiplicity of -1 rises from 8 to 9. In all other (non-isomorphic) examples the difference of the dimension is three and the multiplicity of the roots $-\frac{4}{3}, -1, -\frac{2}{3}$ rises by 1.

Moreover we see that the roots of the Bernstein-Sato polynomials fulfil the exponent condition 8.1.4 and the multiplicity of the root -1 equals the dimension of the maximal torus T, hence the condition $(C3)$ of the invariant subspace conjecture is satisfied. Hence we conclude

Corollary 8.4.7. The pairs $(E_6, \mathbf{d}_i)$, $i = 1, \ldots, 6$, fulfil the exponent condition 8.1.4 and the invariant subspace conjecture 8.1.1.
In particular the corresponding hypergeometric system $G_1(*D)$ can be reduced from a GKZ-system.

We remark that the quiver E_6 supports different quivers by changing the orientation of the arrows. In the case of A_n and D_n the discriminants are isomorphic, but this is not the case for E_6. For example there are non isomorphic discriminants for $\mathbf{d}_6$. However we know from calculations that the spectrum of their generic hyperplane sections does not depend on the orientation.

8.5 Semi-invariants and spectrum for E_7

We consider the Dynkin quiver E_7 with the following ordering of nodes:

$$\begin{array}{ccccccccccc} 1 & \overset{\alpha}{\text{—}} & 2 & \overset{\beta}{\text{—}} & 3 & \overset{\gamma}{\text{—}} & 4 & \overset{\delta}{\text{—}} & 5 & \overset{\epsilon}{\text{—}} & 6 \\ & & & & \zeta\,\vert & & & & & & \\ & & & & 7 & & & & & & . \end{array}$$

By [Bou81] p.264 the positive sincere roots of E_7 are

$\mathbf{d}^1: \begin{smallmatrix}1&1&2&1&1&1\\ &&1&&&\end{smallmatrix}$	$\mathbf{d}^2: \begin{smallmatrix}1&2&2&1&1&1\\ &&1&&&\end{smallmatrix}$
$\mathbf{d}^3: \begin{smallmatrix}1&1&2&2&1&1\\ &&1&&&\end{smallmatrix}$	$\mathbf{d}^4: \begin{smallmatrix}1&2&2&2&1&1\\ &&1&&&\end{smallmatrix}$
$\mathbf{d}^5: \begin{smallmatrix}1&1&2&2&2&1\\ &&1&&&\end{smallmatrix}$	$\mathbf{d}^6: \begin{smallmatrix}1&2&2&2&2&1\\ &&1&&&\end{smallmatrix}$
$\mathbf{d}^7: \begin{smallmatrix}1&2&3&2&1&1\\ &&1&&&\end{smallmatrix}$	$\mathbf{d}^8: \begin{smallmatrix}1&2&3&2&2&1\\ &&1&&&\end{smallmatrix}$
$\mathbf{d}^9: \begin{smallmatrix}1&2&3&2&1&1\\ &&2&&&\end{smallmatrix}$	$\mathbf{d}^{10}: \begin{smallmatrix}1&2&3&3&2&1\\ &&1&&&\end{smallmatrix}$
$\mathbf{d}^{11}: \begin{smallmatrix}1&2&3&2&2&1\\ &&2&&&\end{smallmatrix}$	$\mathbf{d}^{12}: \begin{smallmatrix}1&2&3&3&2&1\\ &&2&&&\end{smallmatrix}$
$\mathbf{d}^{13}: \begin{smallmatrix}1&2&4&3&2&1\\ &&2&&&\end{smallmatrix}$	$\mathbf{d}^{14}: \begin{smallmatrix}1&3&4&3&2&1\\ &&2&&&\end{smallmatrix}$
$\mathbf{d}^{15}: \begin{smallmatrix}2&3&4&3&2&1\\ &&2&&&\end{smallmatrix}$	

Because the Tits form of each of these roots is equal to one, we have:

Proposition 8.5.1. Every root $\mathbf{d}^i$, $i = 1, \ldots, 15$, from above is a real Schur root and each pair $(\mathrm{PGL}(E_7, \mathbf{d}^i), \mathrm{Rep}(E_7, \mathbf{d}^i))$ is a regular reductive prehomogeneous vector space.

Proof. One easily verifies that we have for $i = 1, \ldots, 15$

$$q_{E_7}(\mathbf{d}^i) = 1.$$

Hence by theorem 7.2.15. each $\mathbf{d}^i$ is a real Schur root and by corollary 7.2.16 each pair $(\mathrm{PGL}(E_7, \mathbf{d}^i), \mathrm{Rep}(E_7, \mathbf{d}^i))$ is a reductive prehomogeneous vector space.
Moreover by theorem 7.2.20 the complement of the open orbit is a divisor, which is by theorem 1.3.2 equivalent to the regularity of the reductive prehomogeneous vector space $(\mathrm{PGL}(E_7, \mathbf{d}^i), \mathrm{Rep}(E_7, \mathbf{d}^i))$. □

Remark 8.5.2. The non sincere positive roots of E_7

0 1 2 1 1 1 1	0 1 2 2 1 1 1	0 1 2 2 2 1 1

can be considered as roots of the Dynkin quiver D_6.

Remark 8.5.3. By Kac's result (7.3.8) there are precisely $6 = (E_7)_0 - 1$ fundamental semi-invariants for each pair $(E_7, \mathbf{d}^i)$, $i = 1, \ldots, 15$.

Remark 8.5.4. The Euler matrix for E_7 with our chosen orientation is

$$E = \begin{pmatrix} 1 & -1 & 0 & 0 & 0 & 0 & 0 \\ 0 & 1 & -1 & 0 & 0 & 0 & 0 \\ 0 & 0 & 1 & -1 & 0 & 0 & 0 \\ 0 & 0 & 0 & 1 & -1 & 0 & 0 \\ 0 & 0 & 0 & 0 & 1 & -1 & 0 \\ 0 & 0 & 0 & 0 & 0 & 1 & 0 \\ 0 & 0 & -1 & 0 & 0 & 0 & 1 \end{pmatrix}.$$

and for a generic representation $N \in \mathrm{Rep}(E_7, \mathbf{e})$ and the general representation M of $\mathrm{Rep}(E_7, \mathbf{d}^i)$ the determinant of the map

$$c_{M,N} = \begin{pmatrix} S_2 a - AS_1 \\ S_3 b - BS_2 \\ S_4 c - CS_3 \\ S_5 d - DS_4 \\ S_6 e - ES_5 \\ S_3 f - FS_7 \end{pmatrix}$$

is a semi-invariant. This follows directly form the diagram

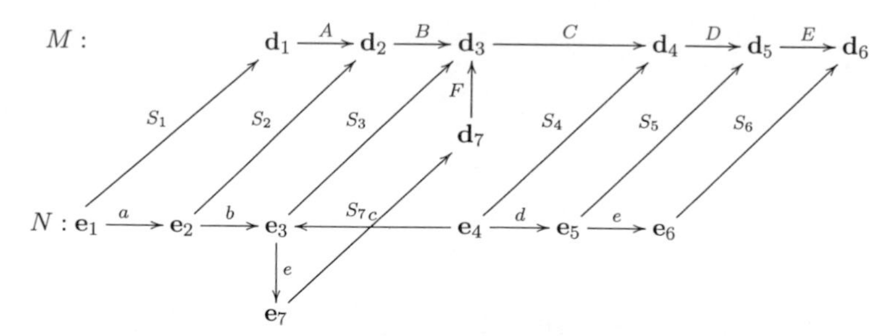

Moreover we can use the Euler matrix to verify or to compute the weight of the semi-invariant c^N using lemma 7.3.2.

In the following we give for each root $\mathbf{d}^i$, $i = 1, \ldots, 15$, the set of fundamental semi-invariants, together with their degree and weights, and the corresponding orthogonal root $\mathbf{e}$.
We give the Bernstein-Sato polynomial of the canonical equation Δ and the Spectrum at $t = 0$ and $t \neq 0$ for the roots $\mathbf{d}^i$, $i = 1, \ldots, 9, 11$, because for the roots $\mathbf{d}^i$, for $i = 10, 12, \ldots, 15$, the Groebner basis computations are out of reach due to memory limitations. I.e. we do not know the spectrum in these cases.

We end each example, where we know the roots of the corresponding Bernstein-Sato polynomial, by giving a representation to solve the invariant subspace conjecture 8.1.1.
Moreover if the degree of Δ is not to big, we also give the equation of the restriction $\Delta_{|W_T}$.

The root $\mathbf{d} = \mathbf{d}^1$:
$$\begin{matrix} \mathbf{1} & \mathbf{1} & \mathbf{2} & \mathbf{1} & \mathbf{1} & \mathbf{1} \\ & & \mathbf{1} & & & \end{matrix}$$

We have $\dim(\mathrm{Rep}(E_7, \mathbf{d})) = 9$ and a dimension vector $\mathbf{e}$ is orthogonal to $\mathbf{d}^1$, if it satisfies the relation $0 = \langle \mathbf{e}, \mathbf{d} \rangle_{E_7} = -e_2 + e_3 + e_6 - e_7$.

Semi-invariant	Degree	Root$^{\perp \mathbf{d}}$	Weight
$h_1 = \det(A)$	1	1 0 0 0 0 0 0	1 -1 0 0 0 0 0
$h_2 = \det(D)$	1	0 0 0 1 0 0 0	0 0 0 1 -1 0 0
$h_3 = \det(E)$	1	0 0 0 0 1 0 0	0 0 0 0 1 -1 0
$h_4 = \det(CB)$	2	0 1 1 0 0 0 0	0 1 0 -1 0 0 0
$h_5 = \det(CF)$	2	0 0 1 0 0 0 1	0 0 0 -1 0 0 1
$h_6 = \det(B\|F)$	2	0 1 1 1 1 1 1	0 1 -1 0 0 0 1
$\Delta = h_1 \ldots h_6$	9		1 1 -1 -1 0 -1 2

The Bernstein-Sato polynomial is

$$b_\Delta(s) = (s + \frac{4}{3})(s + 1)^7(s + \frac{2}{3}).$$

The Spectrum (at $t = 0$ and $t \neq 0$) is

$$(3, 1, 2, 3, 4, 5, 6, 7, 5).$$

If we consider for the invariant subspace conjecture 8.1.1 the representation

$$M = \left(\begin{pmatrix} 1 \end{pmatrix}, \begin{pmatrix} 1 \\ 0 \end{pmatrix}, \begin{pmatrix} 1 & 1 \end{pmatrix}, \begin{pmatrix} 1 \end{pmatrix}, \begin{pmatrix} 1 \end{pmatrix}, \begin{pmatrix} 0 \\ 1 \end{pmatrix} \right),$$

then the restriction of Δ to the subspace $W_T := \overline{T.M}$ is

$$\Delta_{|W_T} = a_{11} b_{11}^2 c_{11} c_{12} d_{11} e_{11} f_{21}^2.$$

The root $\mathbf{d} = \mathbf{d}^2$: $\begin{matrix} \mathbf{1} & \mathbf{2} & \mathbf{2} & \mathbf{1} & \mathbf{1} & \mathbf{1} \\ & & \mathbf{1} & & & \end{matrix}$

We have $\dim(\mathrm{Rep}(E_7, \mathbf{d}^2)) = 12$ and a dimension vector $\mathbf{e}$ is orthogonal to $\mathbf{d}^2$, if it satisfies the relation $0 = \langle \mathbf{e}, \mathbf{d}^2 \rangle_{E_7} = -e_1 + e_3 + e_6 - e_7$.

Semi-invariant	Degree	$\mathrm{Root}^{\perp \mathbf{d}}$	Weight
$h_1 = \det(D)$	1	0 0 0 1 0 0 0	0 0 0 1 -1 0 0
$h_2 = \det(E)$	1	0 0 0 0 1 0 0	0 0 0 0 1 -1 0
$h_3 = \det(CF)$	2	0 0 1 0 0 0 1	0 0 0 -1 0 0 1
$h_4 = \det(B)$	2	0 1 0 0 0 0 0	0 1 -1 0 0 0 0
$h_5 = \det(CBA)$	3	1 1 1 0 0 0 0	1 0 0 -1 0 0 0
$h_6 = \det(BA\|F)$	3	1 1 1 1 1 1 1	1 0 -1 0 0 0 1
$\Delta = h_1 \dots h_6$	12		2 1 -2 -1 0 -1 2

The Bernstein-Sato polynomial is

$$b_\Delta(s) = (s + \frac{4}{3})^2 (s+1)^8 (s + \frac{2}{3})^2.$$

The Spectrum (at $t = 0$ and $t \neq 0$) is

$$(4, 5, 2, \dots, 9, 6, 7).$$

If we consider for the invariant subspace conjecture 8.1.1 the representation

$$M = \left(\begin{pmatrix} 1 \\ 0 \end{pmatrix}, \begin{pmatrix} 1 & 0 \\ 0 & 1 \end{pmatrix}, \begin{pmatrix} 1 & 1 \end{pmatrix}, \begin{pmatrix} 1 \end{pmatrix}, \begin{pmatrix} 1 \end{pmatrix}, \begin{pmatrix} 0 \\ 1 \end{pmatrix} \right),$$

then the restriction of Δ to the subspace $W_T := \overline{T.M}$ is

$$\Delta_{|W_T} = a_{11}^2 b_{11}^3 b_{22} c_{11} c_{12} d_{11} e_{11} f_{21}^2.$$

The root $\mathbf{d} = \mathbf{d}^3$: $\begin{matrix} \mathbf{1} & \mathbf{1} & \mathbf{2} & \mathbf{2} & \mathbf{1} & \mathbf{1} \\ & & \mathbf{1} & & & \end{matrix}$

We have $\dim(\mathrm{Rep}(E_7, \mathbf{d}^3)) = 12$ and a dimension vector $\mathbf{e}$ is orthogonal to $\mathbf{d}^3$, if it satisfies the relation $0 = \langle \mathbf{e}, \mathbf{d}^3 \rangle_{E_7} = -e_2 + e_4 + e_6 - e_7$.

Semi-invariant	Degree	Root$^{\perp \mathbf{d}}$	Weight
$h_1 = \det(A)$	1	1 0 0 0 0 0 0	1 -1 0 0 0 0 0
$h_2 = \det(E)$	1	0 0 0 0 1 0 0	0 0 0 0 1 -1 0
$h_3 = \det(C)$	2	0 0 1 0 0 0 0	0 0 1 -1 0 0 0
$h_4 = \det(B\|F)$	2	0 1 1 1 1 1 1	0 1 -1 0 0 0 1
$h_5 = \det(DCB)$	3	0 1 1 1 0 0 0	0 1 0 0 -1 0 0
$h_6 = \det(DCF)$	3	0 0 1 1 0 0 1	0 0 0 0 -1 0 1
$\Delta = h_1 \ldots h_6$	12		1 1 0 -1 -1 -1 2

The Bernstein-Sato polynomial is

$$b_\Delta(s) = (s + \frac{4}{3})^2 (s+1)^8 (s + \frac{2}{3})^2.$$

The Spectrum (at $t = 0$ and $t \neq 0$) is

$$(4, 5, 2, \ldots, 9, 6, 7).$$

If we consider for the invariant subspace conjecture 8.1.1 the representation

$$M = \left(\begin{pmatrix} 1 \end{pmatrix}, \begin{pmatrix} 1 \\ 0 \end{pmatrix}, \begin{pmatrix} 1 & 0 \\ 0 & 1 \end{pmatrix}, \begin{pmatrix} 1 & 1 \end{pmatrix}, \begin{pmatrix} 1 \end{pmatrix}, \begin{pmatrix} 0 \\ 1 \end{pmatrix} \right),$$

then the restriction of Δ to the subspace $W_T := \overline{T.M}$ is

$$\Delta_{|W_T} = a_{11} b_{11}^2 c_{11}^2 c_{22}^2 d_{11} d_{12} e_{11} f_{21}^2.$$

The root $\mathbf{d} = \mathbf{d}^4$: $\begin{matrix} \mathbf{1} & \mathbf{2} & \mathbf{2} & \mathbf{2} & \mathbf{1} & \mathbf{1} \\ & & \mathbf{1} & & & \end{matrix}$

We have $\dim(\mathrm{Rep}(E_7, \mathbf{d}^4)) = 15$ and a dimension vector $\mathbf{e}$ is orthogonal to $\mathbf{d}^4$, if it satisfies the relation $0 = \langle \mathbf{e}, \mathbf{d}^4 \rangle_{E_7} = -e_1 + e_4 + e_6 - e_7$.

Semi-invariant	Degree	$\mathrm{Root}^{\perp \mathbf{d}}$	Weight
$h_1 = \det(E)$	1	$\begin{matrix} 0 & 0 & 0 & 0 & 1 & 0 \\ & & 0 & & & \end{matrix}$	$\begin{matrix} 0 & 0 & 0 & 0 & 1 & -1 \\ & & 0 & & & \end{matrix}$
$h_2 = \det(B)$	2	$\begin{matrix} 0 & 1 & 0 & 0 & 0 & 0 \\ & & 0 & & & \end{matrix}$	$\begin{matrix} 0 & 1 & -1 & 0 & 0 & 0 \\ & & 0 & & & \end{matrix}$
$h_3 = \det(C)$	2	$\begin{matrix} 0 & 0 & 1 & 0 & 0 & 0 \\ & & 0 & & & \end{matrix}$	$\begin{matrix} 0 & 0 & 1 & -1 & 0 & 0 \\ & & 0 & & & \end{matrix}$
$h_4 = \det(DCF)$	3	$\begin{matrix} 0 & 0 & 1 & 1 & 0 & 0 \\ & & 1 & & & \end{matrix}$	$\begin{matrix} 0 & 0 & 0 & 0 & -1 & 0 \\ & & 1 & & & \end{matrix}$
$h_5 = \det(F\|BA)$	3	$\begin{matrix} 1 & 1 & 1 & 1 & 1 & 1 \\ & & 1 & & & \end{matrix}$	$\begin{matrix} 1 & 0 & -1 & 0 & 0 & 0 \\ & & 1 & & & \end{matrix}$
$h_6 = \det(DCBA)$	4	$\begin{matrix} 1 & 1 & 1 & 1 & 0 & 0 \\ & & 0 & & & \end{matrix}$	$\begin{matrix} 1 & 0 & 0 & 0 & -1 & 0 \\ & & 0 & & & \end{matrix}$
$\Delta = h_1 \dots h_6$	15		$\begin{matrix} 2 & 1 & -1 & -1 & -1 & -1 \\ & & 2 & & & \end{matrix}$

The Bernstein-Sato polynomial is

$$b_\Delta(s) = (s + \frac{4}{3})^3 (s+1)^9 (s + \frac{2}{3})^3.$$

The Spectrum (at $t = 0$ and $t \neq 0$) is

$$(5, 6, 7, 3, \dots, 11, 7, 8, 9).$$

If we consider for the invariant subspace conjecture 8.1.1 the representation

$$M = \left(\begin{pmatrix} 1 \\ 0 \end{pmatrix}, \begin{pmatrix} 1 & 0 \\ 0 & 1 \end{pmatrix}, \begin{pmatrix} 1 & 0 \\ 0 & 1 \end{pmatrix}, \begin{pmatrix} 1 & 1 \end{pmatrix}, \begin{pmatrix} 1 \end{pmatrix}, \begin{pmatrix} 0 \\ 1 \end{pmatrix} \right),$$

then the restriction of Δ to the subspace $W_T := \overline{T.M}$ is

$$\Delta_{|W_T} = a_{11}^2 b_{11}^3 b_{22} c_{11}^2 c_{22}^2 d_{11} d_{12} e_{11} f_{21}^2.$$

The root $\mathbf{d} = \mathbf{d}^5$: $\begin{matrix} 1 & 1 & 2 & 2 & 2 & 1 \\ & & 1 & & & \end{matrix}$

We have $\dim(\mathrm{Rep}(E_7, \mathbf{d}^5)) = 15$ and a dimension vector $\mathbf{e}$ is orthogonal to $\mathbf{d}^5$, if it satisfies the relation $0 = \langle \mathbf{e}, \mathbf{d}^5 \rangle_{E_7} = -e_2 + e_5 + e_6 - e_7$.

Semi-invariant	Degree	Root$^{\perp \mathbf{d}}$	Weight
$h_1 = \det(A)$	1	1 0 0 0 0 0 0	1 -1 0 0 0 0 0
$h_2 = \det(C)$	2	0 0 1 0 0 0 0	0 0 1 -1 0 0 0
$h_3 = \det(D)$	2	0 0 0 1 0 0 0	0 0 0 1 -1 0 0
$h_4 = \det(B\vert F)$	2	0 1 1 1 1 1 1	0 1 -1 0 0 0 1
$h_5 = \det(EDCB)$	4	0 1 1 1 1 0 0	0 1 0 0 0 -1 0
$h_6 = \det(EDCF)$	4	0 0 1 1 1 0 1	0 0 0 0 0 -1 1
$\Delta = h_1 \ldots h_6$	15		1 1 0 0 -1 -2 2

The Bernstein-Sato polynomial is

$$b_\Delta(s) = (s + \frac{4}{3})^3 (s+1)^9 (s + \frac{2}{3})^3.$$

The Spectrum (at $t = 0$ and $t \neq 0$) is

$$(5, 6, 7, 3, \ldots, 11, 7, 8, 9).$$

If we consider for the invariant subspace conjecture 8.1.1 the representation

$$M = \left(\begin{pmatrix} 1 \end{pmatrix}, \begin{pmatrix} 1 \\ 0 \end{pmatrix}, \begin{pmatrix} 1 & 0 \\ 0 & 1 \end{pmatrix}, \begin{pmatrix} 1 & 0 \\ 0 & 1 \end{pmatrix}, \begin{pmatrix} 1 & 1 \end{pmatrix}, \begin{pmatrix} 0 \\ 1 \end{pmatrix} \right),$$

then the restriction of Δ to the subspace $W_T := \overline{T.M}$ is

$$\Delta_{|W_T} = a_{11} b_{11}^2 c_{11}^2 c_{22}^2 d_{11}^2 d_{22}^2 e_{11} e_{12} f_{21}^2.$$

We compare the degrees of the fundamental semi-invariants and conclude:

Proposition 8.5.5. The divisors corresponding to $\mathbf{d}^4$ and $\mathbf{d}^5$ are not isomorphic, but they have the same b–function and the same spectrum.

The root $\mathbf{d} = \mathbf{d}^6$: $\begin{matrix} \mathbf{1} & \mathbf{2} & \mathbf{2} & \mathbf{2} & \mathbf{2} & \mathbf{1} \\ & & \mathbf{1} & & & \end{matrix}$

We have $\dim(\mathrm{Rep}(E_7, \mathbf{d}^6)) = 18$ and a dimension vector $\mathbf{e}$ is orthogonal to $\mathbf{d}^6$, if it satisfies the relation $0 = \langle \mathbf{e}, \mathbf{d}^6 \rangle_{E_7} = -e_1 + e_5 + e_6 - e_7$.

Semi-invariant	Degree	Root$^{\perp \mathbf{d}}$	Weight
$h_1 = \det(B)$	2	0 1 0 0 0 0 0	0 1 -1 0 0 0 0
$h_2 = \det(C)$	2	0 0 1 0 0 0 0	0 0 1 -1 0 0 0
$h_3 = \det(D)$	2	0 0 0 1 0 0 0	0 0 0 1 -1 0 0
$h_4 = \det(F\|BA)$	3	1 1 1 1 1 1 1	1 0 -1 0 0 0 1
$h_5 = \det(EDCF)$	4	0 0 1 1 1 0 1	0 0 0 0 0 -1 1
$h_6 = \det(EDCBA)$	5	1 1 1 1 1 0 0	1 0 0 0 0 -1 0
$\Delta = h_1 \dots h_6$	18		2 1 -1 0 -1 -2 2

The Bernstein-Sato polynomial is

$$b_\Delta(s) = (s + \frac{4}{3})^4 + (s+1)^{10}(s + \frac{2}{3})^4.$$

The Spectrum (at $t = 0$ and $t \neq 0$) is

$$(6, 7, 8, 9, 4, \dots, 13, 8, 9, 10, 11).$$

If we consider for the invariant subspace conjecture 8.1.1 the representation

$$M = \left(\begin{pmatrix} 1 \\ 0 \end{pmatrix}, \begin{pmatrix} 1 & 0 \\ 0 & 1 \end{pmatrix}, \begin{pmatrix} 1 & 0 \\ 0 & 1 \end{pmatrix}, \begin{pmatrix} 1 & 0 \\ 0 & 1 \end{pmatrix}, \begin{pmatrix} 1 & 1 \end{pmatrix}, \begin{pmatrix} 0 \\ 1 \end{pmatrix} \right),$$

then the restriction of Δ to the subspace $W_T := \overline{T.M}$ is

$$\Delta_{|W_T} = a_{11}^2 b_{11}^3 b_{22} c_{11}^2 c_{22}^2 d_{11}^2 d_{22}^2 e_{11} e_{12} f_{21}^2.$$

The root $\mathbf{d} = \mathbf{d}^7$: $\begin{matrix} \mathbf{1} & \mathbf{2} & \mathbf{3} & \mathbf{2} & \mathbf{1} & \mathbf{1} \\ & & \mathbf{1} & & & \end{matrix}$

We have $\dim(\mathrm{Rep}(E_7, \mathbf{d}^7)) = 20$ and a dimension vector $\mathbf{e}$ is orthogonal to $\mathbf{d}^7$, if it satisfies the relation $0 = \langle \mathbf{e}, \mathbf{d}^7 \rangle_{E_7} = -e_1 - e_2 + e_3 + e_4 + e_6 - 2e_7$.

Semi-invariant	Degree	Root$^{\perp \mathbf{d}}$	Weight
$h_1 = \det(E)$	1	0 0 0 0 1 0 0	0 0 0 0 1 -1 0
$h_2 = \det(DCF)$	3	0 0 1 1 0 0 1	0 0 0 0 -1 0 1
$h_3 = \det(CB)$	4	0 1 1 0 0 0 0	0 1 0 -1 0 0 0
$h_4 = \det(DCBA)$	4	1 1 1 1 0 0 0	1 0 0 0 -1 0 0
$h_5 = \det(B\|F)$	3	0 1 1 1 1 1 1	0 1 -1 0 0 0 1
$h_6 = \det(CBA\|CF)$	5	1 1 2 1 1 1 1	1 0 0 -1 0 0 1
$\Delta = h_1 \dots h_6$	20		2 2 -1 -2 -1 -1 3

The Bernstein-Sato polynomial is

$$b_\Delta(s) = (s + \frac{7}{5})(s + \frac{4}{3})^3(s + \frac{6}{5})(s + 1)^{10}(s + \frac{4}{5})(s + \frac{2}{3})^3(s + \frac{3}{5}).$$

The Spectrum (at $t = 0$ and $t \neq 0$) is

$$(8, \frac{23}{3}, \frac{26}{3}, \frac{29}{3}, 8, 5, \dots, 14, 11, \frac{28}{3}, \frac{31}{3}, \frac{34}{3}, 11).$$

If we consider for the invariant subspace conjecture 8.1.1 the representation

$$M = \left(\begin{pmatrix} 1 \\ 0 \end{pmatrix}, \begin{pmatrix} 1 & 0 \\ 0 & 1 \\ 1 & 1 \end{pmatrix}, \begin{pmatrix} 1 & 0 & 1 \\ 0 & 1 & 1 \end{pmatrix}, \begin{pmatrix} 0 & 1 \end{pmatrix}, \begin{pmatrix} 1 \end{pmatrix}, \begin{pmatrix} 1 \\ 1 \\ 1 \end{pmatrix} \right),$$

then the restriction of Δ to the subspace $W_T := \overline{T.M}$ is a monomial of degree 20 in 10 variables.

The root $\mathbf{d} = \mathbf{d}^8$: $\begin{matrix} \mathbf{1} & \mathbf{2} & \mathbf{3} & \mathbf{2} & \mathbf{2} & \mathbf{1} \\ & & \mathbf{1} & & & \end{matrix}$

We have $\dim(\text{Rep}(E_7, \mathbf{d}^8)) = 23$ and a dimension vector $\mathbf{e}$ is orthogonal to $\mathbf{d}^8$, if it satisfies the relation $0 = \langle \mathbf{e}, \mathbf{d}^8 \rangle_{E_7} = -e_1 - e_2 + e_3 + e_5 + e_6 - 2e_7$.

Semi-invariant	Degree	Root$^{\perp \mathbf{d}}$	Weight
$h_1 = \det(D)$	2	0 0 0 1 0 0 0	0 0 0 1 -1 0 0
$h_2 = \det(B\|F)$	3	0 1 1 1 1 1 1	0 1 -1 0 0 0 1
$h_3 = \det(CB)$	4	0 1 1 0 0 0 0	0 1 0 -1 0 0 0
$h_4 = \det(EDCF)$	4	0 0 1 1 1 0 1	0 0 0 0 0 -1 1
$h_5 = \det(EDCBA)$	5	1 1 1 1 1 0 0	1 0 0 0 0 -1 0
$h_6 = \det(CBA\|CF)$	5	1 1 2 1 1 1 1	1 0 0 -1 0 0 1
$\Delta = h_1 \dots h_6$	23		2 2 -1 -1 -1 -2 3

The Bernstein-Sato polynomial is

$$b_\Delta(s) = (s + \frac{7}{5})(s + \frac{4}{3})^4(s + \frac{6}{5})(s + 1)^{11}(s + \frac{4}{5})(s + \frac{2}{3})^4(s + \frac{3}{5}).$$

The Spectrum (at $t = 0$ and $t \neq 0$) is

$$(\frac{46}{5}, \frac{26}{3}, \frac{29}{3}, \frac{32}{3}, \frac{35}{3}, \frac{48}{5}, 6, \dots, 16, \frac{62}{5}, \frac{31}{3}, \frac{34}{3}, \frac{37}{3}, \frac{40}{3}, \frac{64}{5}).$$

If we consider for the invariant subspace conjecture 8.1.1 the representation

$$M = \left(\begin{pmatrix} 1 \\ 0 \end{pmatrix}, \begin{pmatrix} 1 & 0 \\ 0 & 1 \\ 1 & 1 \end{pmatrix}, \begin{pmatrix} 1 & 0 & 1 \\ 0 & 1 & 1 \end{pmatrix}, \begin{pmatrix} 1 & 0 \\ 0 & 1 \end{pmatrix}, \begin{pmatrix} 0 & 1 \end{pmatrix}, \begin{pmatrix} 1 \\ 1 \\ 1 \end{pmatrix} \right),$$

then the restriction of Δ to the subspace $W_T := \overline{T.M}$ is a monomial in 11 variables of degree 23.

The root $\mathbf{d} = \mathbf{d}^9$: $\begin{matrix} \mathbf{1} & \mathbf{2} & \mathbf{3} & \mathbf{2} & \mathbf{1} & \mathbf{1} \\ & & \mathbf{2} & & & \end{matrix}$

We have $\dim(\mathrm{Rep}(E_7, \mathbf{d}^9)) = 23$ and a dimension vector $\mathbf{e}$ is orthogonal to $\mathbf{d}^9$, if it satisfies the relation $0 = \langle \mathbf{e}, \mathbf{d}^9 \rangle_{E_7} = -e_1 - e_2 + e_3 + e_4 + e_6 - e_7$.

Semi-invariant	Degree	$\mathrm{Root}^{\perp \mathbf{d}}$	Weight
$h_1 = \det(E)$	1	0 0 0 0 1 0 0	0 0 0 0 1 -1 0
$h_2 = \det(CB)$	4	0 1 1 0 0 0 0	0 1 0 -1 0 0 0
$h_3 = \det(CF)$	4	0 0 1 0 0 0 1	0 0 0 -1 0 0 1
$h_4 = \det(DCBA)$	4	1 1 1 1 0 0 0	1 0 0 0 -1 0 0
$h_5 = \det(BA\|F)$	4	1 1 1 1 1 1 1	1 0 -1 0 0 0 1
$h_6 = \det(\Gamma)$	6	0 1 1 1 0 0 1	0 1 -1 0 -1 0 1
$\Delta = h_1 \dots h_6$	23		2 2 -2 -2 -1 -1 3

where

$$\Gamma = \begin{pmatrix} -B & \mathbf{1}_3 & 0 & 0 \\ 0 & -C & \mathbf{1}_2 & 0 \\ 0 & 0 & -D & 0 \\ 0 & \mathbf{1}_3 & 0 & -F \end{pmatrix}$$

The Bernstein-Sato polynomial is

$$b_\Delta(s) = (s + \frac{7}{5})(s + \frac{4}{3})^4(s + \frac{6}{5})(s + 1)^{11}(s + \frac{4}{5})(s + \frac{2}{3})^4(s + \frac{3}{5}).$$

The Spectrum (at $t = 0$ and $t \neq 0$) is

$$(\frac{46}{5}, \frac{26}{3}, \frac{29}{3}, \frac{32}{3}, \frac{35}{3}, \frac{48}{5}, 6, \dots, 16, \frac{62}{5}, \frac{31}{3}, \frac{34}{3}, \frac{37}{3}, \frac{40}{3}, \frac{64}{5}).$$

We compare the degrees of the fundamental semi-invariants and conclude:

Proposition 8.5.6. The divisors corresponding to $\mathbf{d}^8$ and $\mathbf{d}^9$ are not isomorphic, but they have the same $b-$function and the same spectrum.

If we consider for the invariant subspace conjecture 8.1.1 the representation

$$M = \left(\begin{pmatrix} 1 \\ 0 \end{pmatrix}, \begin{pmatrix} 1 & 0 \\ 0 & 1 \\ 1 & 1 \end{pmatrix}, \begin{pmatrix} 1 & 0 & 1 \\ 0 & 1 & 1 \end{pmatrix}, \begin{pmatrix} 0 & 1 \end{pmatrix}, \begin{pmatrix} 1 \end{pmatrix}, \begin{pmatrix} 1 & 0 \\ 1 & 1 \\ 0 & 2 \end{pmatrix} \right),$$

then the restriction of Δ to the subspace $W_T := \overline{T.M}$ is a monomial in 11 variables of degree 23.

The root $\mathbf{d} = \mathbf{d}^{10}$: $\begin{matrix} \mathbf{1} & \mathbf{2} & \mathbf{3} & \mathbf{3} & \mathbf{2} & \mathbf{1} \\ & & \mathbf{1} & & & \end{matrix}$

We have $\dim(\mathrm{Rep}(E_7, \mathbf{d}^{10})) = 28$ and a dimension vector $\mathbf{e}$ is orthogonal to $\mathbf{d}^{10}$, if it satisfies the relation $0 = \langle \mathbf{e}, \mathbf{d}^{10} \rangle_{E_7} = -e_1 - e_2 + e_4 + e_5 + e_6 - 2e_7$.

Semi-invariant	Degree	$\mathrm{Root}^{\perp \mathbf{d}}$	Weight
$h_1 = \det(C)$	3	0 0 1 0 0 0 0	0 0 1 -1 0 0 0
$h_2 = \det(EDCF)$	4	0 0 1 1 1 0 1	0 0 0 0 0 -1 1
$h_3 = \det(EDCBA)$	5	1 1 1 1 1 0 0	1 0 0 0 0 -1 0
$h_4 = \det(DCB)$	6	0 1 1 1 0 0 0	0 1 0 0 -1 0 0
$h_5 = \det(B\|F)$	3	0 1 1 1 1 1 1	0 1 -1 0 0 0 1
$h_6 = \det(DCF\|DCBA)$	7	1 1 2 2 1 1 1	1 0 0 0 -1 0 1
$\Delta = h_1 \dots h_6$	28		2 2 0 -1 -2 -2 3

Remark 8.5.7. The Groebner basis computations could not be finished due to memory limitations.

The root $\mathbf{d} = \mathbf{d}^{11}$: $\begin{matrix} \mathbf{1} & \mathbf{2} & \mathbf{3} & \mathbf{2} & \mathbf{2} & \mathbf{1} \\ & & \mathbf{2} & & & \end{matrix}$

We have $\dim(\mathrm{Rep}(E_7, \mathbf{d}^{11})) = 26$ and a dimension vector $\mathbf{e}$ is orthogonal to $\mathbf{d}^{11}$, if it satisfies the relation $0 = \langle \mathbf{e}, \mathbf{d}^{11} \rangle_{E_7} = -e_1 - e_2 + e_3 + e_5 + e_6 - e_7$.

Semi-invariant	Degree	Root$^{\perp \mathbf{d}}$	Weight
$h_1 = \det(D)$	2	0 0 0 1 0 0 0	0 0 0 1 -1 0 0
$h_2 = \det(CB)$	4	0 1 1 0 0 0 0	0 1 0 -1 0 0 0
$h_3 = \det(CF)$	4	0 0 1 0 0 0 1	0 0 0 -1 0 0 1
$h_4 = \det(BA\|F)$	4	1 1 1 1 1 1 1	1 0 -1 0 0 0 1
$h_5 = \det(EDCBA)$	5	1 1 1 1 1 0 0	1 0 0 0 0 -1 0
$h_6 = \det(\Gamma)$	7	0 1 1 1 1 0 1	0 1 -1 0 0 -1 1
$\Delta = h_1 \dots h_6$	26		2 2 -2 -1 -1 -2 3

where

$$\Gamma = \begin{pmatrix} -B & \mathbf{1}_3 & 0 & 0 & 0 \\ 0 & -C & \mathbf{1}_2 & 0 & 0 \\ 0 & 0 & -D & \mathbf{1}_2 & 0 \\ 0 & 0 & 0 & -E & 0 \\ 0 & \mathbf{1}_3 & 0 & 0 & -F \end{pmatrix}$$

The Bernstein-Sato polynomial is

$$b_\Delta(s) = (s+\frac{7}{5})(s+\frac{4}{3})^5(s+\frac{6}{5})(s+1)^{12}(s+\frac{4}{5})(s+\frac{2}{3})^5(s+\frac{3}{5}).$$

The Spectrum (at $t = 0$ and $t \neq 0$) is

$$(\frac{52}{5}, \frac{29}{3}, \frac{32}{3}, \frac{35}{3}, \frac{38}{3}, \frac{41}{3}, \frac{56}{5}, 7, \dots, 18, \frac{69}{5}, \frac{34}{3}, \frac{37}{3}, \frac{40}{3}, \frac{43}{3}, \frac{46}{3}, \frac{73}{5}).$$

If we consider for the invariant subspace conjecture 8.1.1 the representation

$$M = \left(\begin{pmatrix} 1 \\ 0 \end{pmatrix}, \begin{pmatrix} 1 & 0 \\ 0 & 1 \\ 1 & 1 \end{pmatrix}, \begin{pmatrix} 1 & 0 & 1 \\ 0 & 1 & 1 \end{pmatrix}, \begin{pmatrix} 1 & 0 \\ 0 & 1 \end{pmatrix}, \begin{pmatrix} 0 & 1 \end{pmatrix}, \begin{pmatrix} 1 & 0 \\ 1 & 1 \\ 0 & 0 \end{pmatrix} \right),$$

then the restriction of Δ to the subspace $W_T := \overline{T.M}$ is a monomial in 12 variables of degree 26.

The root $\mathbf{d}^{12}$: $\begin{matrix} \mathbf{1} & \mathbf{2} & \mathbf{3} & \mathbf{3} & \mathbf{2} & \mathbf{1} \\ & & \mathbf{2} & & & \end{matrix}$

We have $\dim(\mathrm{Rep}(E_7, \mathbf{d}^{12})) = 31$ and a dimension vector $\mathbf{e}$ is orthogonal, if it satisfies the relation $0 = \langle \mathbf{e}, \mathbf{d}^{12} \rangle_{E_7} = -e_1 - e_2 + e_4 + e_5 + e_6 - e_7$.

Semi-invariant	Degree	Root$^{\perp\mathbf{d}}$	Weight
$h_1 = \det(C)$	3	0 0 1 0 0 0 0	0 0 1 -1 0 0 0
$h_2 = \det(BA\|F)$	4	1 1 1 1 1 1 1	1 0 -1 0 0 0 1
$h_3 = \det(EDCBA)$	5	1 1 1 1 1 0 0	1 0 0 0 0 -1 0
$h_4 = \det(DCB)$	6	0 1 1 1 0 0 0	0 1 0 0 -1 0 0
$h_5 = \det(DCF)$	6	0 0 1 1 0 0 1	0 0 0 0 -1 0 1
$h_6 = \det(\Gamma)$	7	0 1 1 1 1 0 1	0 1 -1 0 0 -1 1
$\Delta = h_1 \dots h_6$	31		2 2 -1 -1 -2 -2 3

where

$$\Gamma = \begin{pmatrix} -B & \mathbf{1}_3 & 0 & 0 & 0 \\ 0 & -C & \mathbf{1}_2 & 0 & 0 \\ 0 & 0 & -D & \mathbf{1}_2 & 0 \\ 0 & 0 & 0 & -E & 0 \\ 0 & \mathbf{1}_3 & 0 & 0 & -F \end{pmatrix}$$

Remark 8.5.8. The Groebner basis computations could not be finished due to memory limitations.

The root $\mathbf{d} = \mathbf{d}^{13}$: $\begin{array}{cccccc} \mathbf{1} & \mathbf{2} & \mathbf{4} & \mathbf{3} & \mathbf{2} & \mathbf{1} \\ & & \mathbf{2} & & & \end{array}$

We have $\dim(\text{Rep}(E_7, \mathbf{d}^{13})) = 38$ and a dimension vector $\mathbf{e}$ is orthogonal, if it satisfies the relation $0 = \langle \mathbf{e}, \mathbf{d}^{13} \rangle_{E_7} = -e_1 - 2e_2 + e_3 + e_4 + e_5 + e_6 - 2e_7$.

Semi-invariant	Degree	$\text{Root}^{\perp \mathbf{d}}$	Weight
$h_1 = \det(B\|F)$	4	$\begin{array}{cccccc} 0 & 1 & 1 & 1 & 1 & 1 \\ & & 1 & & & \end{array}$	$\begin{array}{cccccc} 0 & 1 & -1 & 0 & 0 & 0 \\ & & 1 & & & \end{array}$
$h_2 = \det(EDCBA)$	5	$\begin{array}{cccccc} 1 & 1 & 1 & 1 & 1 & 0 \\ & & 0 & & & \end{array}$	$\begin{array}{cccccc} 1 & 0 & 0 & 0 & 0 & -1 \\ & & 0 & & & \end{array}$
$h_3 = \det(DCB)$	6	$\begin{array}{cccccc} 0 & 1 & 1 & 1 & 0 & 0 \\ & & 0 & & & \end{array}$	$\begin{array}{cccccc} 0 & 1 & 0 & 0 & -1 & 0 \\ & & 0 & & & \end{array}$
$h_4 = \det(DCF)$	6	$\begin{array}{cccccc} 0 & 0 & 1 & 1 & 0 & 0 \\ & & 1 & & & \end{array}$	$\begin{array}{cccccc} 0 & 0 & 0 & 0 & -1 & 0 \\ & & 1 & & & \end{array}$
$h_5 = \det(CBA\|CF)$	7	$\begin{array}{cccccc} 1 & 1 & 2 & 1 & 1 & 1 \\ & & 1 & & & \end{array}$	$\begin{array}{cccccc} 1 & 0 & 0 & -1 & 0 & 0 \\ & & 1 & & & \end{array}$
$h_6 = (\Gamma)$	10	$\begin{array}{cccccc} 0 & 1 & 2 & 1 & 1 & 0 \\ & & 1 & & & \end{array}$	$\begin{array}{cccccc} 0 & 1 & 0 & -1 & 0 & -1 \\ & & 1 & & & \end{array}$
$\Delta = h_1 \dots h_6$	38		$\begin{array}{cccccc} 2 & 3 & -1 & -2 & -2 & -2 \\ & & 4 & & & \end{array}$

where

$$\Gamma = \begin{pmatrix} -B & \mathbf{1}_4 & 0 & 0 & 0 & 0 \\ 0 & -C & 0 & 0 & 0 & 0 \\ 0 & 0 & -C & \mathbf{1}_3 & 0 & 0 \\ 0 & 0 & 0 & -D & \mathbf{1}_2 & 0 \\ 0 & 0 & 0 & 0 & -E & 0 \\ 0 & \mathbf{1}_4 & \mathbf{1}_4 & 0 & 0 & -F \end{pmatrix}$$

Remark 8.5.9. The Groebner basis computations could not be finished due to memory limitations.

The root $\mathbf{d}^{14}$: $\begin{array}{cccccc} \mathbf{1} & \mathbf{3} & \mathbf{4} & \mathbf{3} & \mathbf{2} & \mathbf{1} \\ & & \mathbf{2} & & & \end{array}$

We have $\dim(\mathrm{Rep}(E_7, \mathbf{d}^{14})) = 43$ and a dimension vector $\mathbf{e}$ is orthogonal, if it satisfies the relation $0 = \langle \mathbf{e}, \mathbf{d}^{14} \rangle_{E_7} = -2e_1 - e_2 + e_3 + e_4 + e_5 + e_6 - 2e_7$.

Semi-invariant	Degree	Root$^{\perp \mathbf{d}}$	Weight
$h_1 = \det(\Gamma_1)$	8	0 1 1 1 1 0 1	0 1 -1 0 0 -1 1
$h_2 = \det(EDCBA)$	5	1 1 1 1 1 0 0	1 0 0 0 0 -1 0
$h_3 = \det(CB)$	6	0 1 1 0 0 0 0	0 1 0 -1 0 0 0
$h_4 = \det(DCF)$	6	0 0 1 1 0 0 1	0 0 0 0 -1 0 1
$h_5 = \det(CBA\vert CF)$	7	1 1 2 1 1 1 1	1 0 0 -1 0 0 1
$h_6 = \det(\Gamma 2)$	11	1 2 2 2 1 1 1	1 1 -1 0 -1 0 1
$\Delta = h_1 \dots h_6$	43		3 3 -2 -2 -2 -2 4

where

$$\Gamma_1 = \begin{pmatrix} -B & \mathbf{1}_4 & 0 & 0 & 0 \\ 0 & -C & \mathbf{1}_3 & 0 & 0 \\ 0 & 0 & -D & \mathbf{1}_2 & 0 \\ 0 & 0 & 0 & -E & 0 \\ 0 & \mathbf{1}_4 & 0 & 0 & -F \end{pmatrix}$$

and

$$\Gamma_2 = \begin{pmatrix} -A & \mathbf{1}_3 & 0 & 0 & 0 & 0 & & & \\ 0 & B & 0 & \mathbf{1}_4 & 0 & 0 & 0 & 0 & 0 \\ 0 & 0 & -B & 0 & \mathbf{1}_4 & 0 & 0 & 0 & \\ 0 & 0 & 0 & -C & 0 & \mathbf{1}_3 & 0 & 0 & \\ 0 & 0 & 0 & 0 & -C & 0 & \mathbf{1}_3 & 0 & 0 \\ 0 & 0 & 0 & 0 & 0 & -D & 0 & 0 & 0 \\ 0 & 0 & 0 & 0 & 0 & 0 & -D & \mathbf{1}_2 & 0 \\ 0 & 0 & 0 & 0 & 0 & 0 & 0 & -E & \mathbf{1}_1 \\ 0 & 0 & 0 & \mathbf{1}_4 & \mathbf{1}_4 & 0 & 0 & 0 & -F \end{pmatrix}.$$

Remark 8.5.10. The Groebner basis computations could not be finished due to memory limitations.

The root $\mathbf{d} = \mathbf{d}^{15}$: $\begin{matrix} \mathbf{2} & \mathbf{3} & \mathbf{4} & \mathbf{3} & \mathbf{2} & \mathbf{1} \\ & & \mathbf{2} & & & \end{matrix}$

We have $\dim(\mathrm{Rep}(E_7, \mathbf{d}^{15})) = 46$ and a dimension vector $\mathbf{e}$ is orthogonal, if it satisfies the relation $0 = \langle \mathbf{e}, \mathbf{d}^{15} \rangle_{E_7} = -e_1 - e_2 + e_3 + e_4 + e_5 + e_6 - 2e_7$.

Semi-invariant	Degree	Root$^{\perp \mathbf{d}}$	Weight
$h_1 = \det(\Gamma_1)$	8	0 1 1 1 1 0 1	0 1 -1 0 0 -1 1
$h_2 = \det(DCF)$	6	0 0 1 1 0 0 1	0 0 0 0 -1 0 1
$h_3 = \det(CB)$	6	0 1 1 0 0 0 0	0 1 0 -1 0 0 0
$h_4 = \det(BA\|F)$	6	1 1 1 1 1 1 1	1 0 -1 0 0 0 1
$h_5 = \det(DCBA)$	8	1 1 1 1 0 0	1 0 0 0 -1 0 0
$h_6 = \det(\Gamma_2)$	12	1 1 2 1 1 0 1	1 0 0 -1 0 -1 1
$\Delta = h_1 \dots h_6$	46		3 2 -2 -2 -2 -2 4

where

$$\Gamma_1 = \begin{pmatrix} -B & \mathbf{1}_4 & 0 & 0 & 0 \\ 0 & -C & \mathbf{1}_3 & 0 & 0 \\ 0 & 0 & -D & \mathbf{1}_2 & 0 \\ 0 & 0 & 0 & -E & 0 \\ 0 & \mathbf{1}_4 & 0 & 0 & -F \end{pmatrix}$$

and

$$\Gamma_2 = \begin{pmatrix} -A & \mathbf{1}_3 & 0 & 0 & 0 & 0 & 0 \\ 0 & -B & \mathbf{1}_4 & 0 & 0 & 0 & 0 \\ 0 & 0 & -C & 0 & 0 & 0 & 0 \\ 0 & 0 & 0 & -C & \mathbf{1}_3 & 0 & 0 \\ 0 & 0 & 0 & 0 & -D & \mathbf{1}_2 & 0 \\ 0 & 0 & 0 & 0 & 0 & -E & 0 \\ 0 & 0 & \mathbf{1}_4 & \mathbf{1}_4 & 0 & 0 & -F \end{pmatrix}$$

Remark 8.5.11. The Groebner basis computations could not be finished due to memory limitations.

Evaluation of E_7

Let us summarize the preceding discussion:
First we remark that we were not able to compute the spectrum of E_7 for the roots $\mathbf{d}^i$, $i = 10, 12, \ldots, 15$, as the Groebner basis computation could not be finished due to memory limitations. Hence any following general statements holds for all cases except $\mathbf{d}^i$, $i = 10, 12, \ldots, 15$.

The spectrum of E_7 consists of positive rational numbers and in particular in the cases $\mathbf{d}^i$, $i = 1, \ldots, 6$ they are integers. Moreover they fulfil the symmetry $\nu_i + \nu_{n+1-i} = \deg(\Delta) - 1$.

We see until $\mathbf{d}^7$ that only the rational numbers $-\frac{4}{3}, -1, \frac{2}{3}$ appear as roots of the Bernstein-Sato polynomial $b_\Delta(s)$ and their multiplicity rises, when the degree of $b_\Delta(s)$ is rising. With root $\mathbf{d}^7$ the roots $-\frac{7}{5}, -\frac{6}{5}, -\frac{4}{5}, -\frac{3}{5}$ appears with multiplicity one. When moving on to the higher roots $\mathbf{d}^i$, $i = 8, 9, 11$ the multiplicity of the roots $-\frac{4}{3}, -1, -\frac{2}{3}$ increases and the multiplicity of the roots $-\frac{7}{5}, -\frac{6}{5}, -\frac{4}{5}, -\frac{3}{5}$ remains one.

Moreover we see that the roots of $b_\Delta(s)$ fulfil the exponent condition 8.1.4 and the multiplicity of the root -1 is equal to the dimension of the maximal torus T. Hence the condition $(C3)$ of the invariant subspace conjecture is true. We conclude

Corollary 8.5.12. The pairs $(E_7, \mathbf{d}^i)$, $i = 1, \ldots, 9, 11$, fulfil the exponent condition 8.1.4, so the corresponding hypergeometric systems $G_1(*D)$ can be reduced from a $GKZ-$system.
Furthermore these pairs fulfil the the invariant subspace conjecture 8.1.1.

We remark that the pairs $(\mathbf{d}^4, \mathbf{d}^5)$ and $(\mathbf{d}^8, \mathbf{d}^9)$ give rise to non isomorphic discriminants, but they have the same spectrum, the same Bernstein-Sato polynomial and the dimensions of the corresponding representation spaces are equal.
We observed in calculations that for a fixed Dynkin quiver and a fixed real Schur root, the roots of the Bernstein-Sato polynomial and the spectral numbers are independent of the chosen orientation. Hence we conjecture:

Conjecture 8.5.13. The roots of the Bernstein-Sato polynomial and the spectral numbers depend only on the Dynkin quiver Q and the real Schur root $\mathbf{d}$. They are independent of the chosen orientation of the arrows.

Remark 8.5.14. The difference $\dim(\mathrm{Rep}(E_7, \mathbf{d}^{i+1})) - \dim(\mathrm{Rep}(E_7, \mathbf{d}^i))$ is equally distributed in a certain sense to the multiplicities of the roots of the Bernstein-Sato polynomial.

For example $\dim(\operatorname{Rep}(E_7, \mathbf{d}^2)) - \dim(\operatorname{Rep}(E_7, \mathbf{d}^1)) = 3$ and the multiplicity of each root $-\frac{4}{3}, -1, -\frac{2}{3}$ rises by 1, while the multiplicity of the roots $-\frac{7}{5}, -\frac{6}{5}, -\frac{4}{5}, -\frac{3}{5}$ remain zero.
On the other hand $\dim(\operatorname{Rep}(E_7, \mathbf{d}^7)) - \dim(\operatorname{Rep}(E_7, \mathbf{d}^5)) = 5$) and the multiplicity of each root $-\frac{7}{5}, -\frac{6}{5}, -1, -\frac{4}{5}, -\frac{3}{5}$ rises by 1, while the multiplicity of the roots $-\frac{4}{3}, -\frac{2}{3}$ remain three. For the pair $(\mathbf{d}^7, \mathbf{d}^6)$ we have

$$\dim(\operatorname{Rep}(E_7, \mathbf{d}^7)) - \dim(\operatorname{Rep}(E_7, \mathbf{d}^6)) = 2 = 4 - 2$$

and we note that the multiplicity of the two roots $-\frac{4}{3}, -\frac{2}{3}$ decreases by 1 and the multiplicity of the four roots $-\frac{7}{5}, -\frac{6}{5}, -\frac{4}{5}, -\frac{3}{5}$ increases by 1. Hence we conjecture:

Conjecture 8.5.15. The Bernstein-Sato polynomials and the spectral numbers for the missing roots of E_7 are given in the following tables:

root	Bernstein-Sato polynomial $b_\Delta(s)$
$\mathbf{d}^{10}$	$(s+\frac{7}{5})^2(s+\frac{4}{3})^4(s+\frac{6}{5})^2(s+1)^{12}(s+\frac{4}{5})^2(s+\frac{2}{3})^4(s+\frac{3}{5})^2$
$\mathbf{d}^{12}$	$(s+\frac{7}{5})^2(s+\frac{4}{3})^5(s+\frac{6}{5})^2(s+1)^{13}(s+\frac{4}{5})^2(s+\frac{2}{3})^5(s+\frac{3}{5})^2$
$\mathbf{d}^{13}$	$(s+\frac{7}{5})^3(s+\frac{4}{3})^6(s+\frac{6}{5})^3(s+1)^{14}(s+\frac{4}{5})^3(s+\frac{2}{3})^6(s+\frac{3}{5})^3$
$\mathbf{d}^{14}$	$(s+\frac{7}{5})^4(s+\frac{4}{3})^6(s+\frac{6}{5})^4(s+1)^{15}(s+\frac{4}{5})^4(s+\frac{2}{3})^6(s+\frac{3}{5})^4$
$\mathbf{d}^{15}$	$(s+\frac{7}{5})^4(s+\frac{4}{3})^7(s+\frac{6}{5})^4(s+1)^{16}(s+\frac{4}{5})^4(s+\frac{2}{3})^7(s+\frac{3}{5})^4$

root	Spectrum (at $t=0$ and $t \neq 0$)
$\mathbf{d}^{10}$	$(\frac{56}{5}, \frac{61}{5}, \frac{34}{3}, \frac{37}{3}, \frac{40}{3}, \frac{43}{3}, \frac{58}{5}, \frac{63}{5}, 8, \ldots, 19, \frac{72}{5}, \frac{77}{5}, \frac{38}{3}, \frac{41}{3}, \frac{44}{3}, \frac{47}{3}, \frac{74}{5}, \frac{79}{5})$
$\mathbf{d}^{12}$	$(\frac{62}{5}, \frac{67}{5}, \frac{37}{3}, \frac{40}{3}, \ldots, \frac{49}{3}, \frac{66}{5}, \frac{71}{5}, 9, \ldots, 21, \frac{79}{5}, \frac{84}{5}, \frac{41}{3}, \ldots, \frac{53}{3}, \frac{83}{5}, \frac{88}{5})$
$\mathbf{d}^{13}$	$(\frac{76}{5}, \frac{81}{5}, \frac{86}{5}, \frac{47}{3}, \ldots, \frac{62}{3}, \frac{83}{5}, \frac{88}{5}, \frac{93}{5}, 12, \ldots, 25, \frac{92}{5}, \frac{97}{5}, \frac{102}{5}, \frac{49}{3}, \ldots, \frac{64}{3}, \frac{99}{5}, \frac{104}{5}, \frac{109}{5})$
$\mathbf{d}^{14}$	$(\frac{86}{5}, \ldots, \frac{101}{5}, \frac{55}{3}, \ldots, \frac{70}{3}, \frac{93}{5}, \ldots, \frac{108}{5}, 14, \ldots, 28, \frac{102}{5}, \ldots, \frac{117}{5}, \frac{56}{3}, \ldots, \frac{71}{3}, \frac{109}{5}, \ldots, \frac{124}{5})$
$\mathbf{d}^{15}$	$(\frac{92}{5}, \ldots, \frac{107}{5}, \frac{58}{3}, \ldots, \frac{76}{3}, \frac{101}{5}, \ldots, \frac{116}{5}, 15, \ldots, 30, \frac{109}{5}, \ldots, \frac{124}{5}, \frac{59}{3}, \ldots, \frac{77}{3}, \frac{118}{5}, \ldots, \frac{133}{5})$

Corollary 8.5.16. If the above conjecture holds true, then the serie E_7 fulfils the invariant subspace conjecture 8.1.1 and the exponent condition 8.1.4.

Proof. The exponent condition 8.1.4 follows by inspection. For the invariant subspace conjecture choose an generic representation M, define $W_T := \overline{T.M}$ and compute the restrictions $(h_i)_{W_T}$ of the fundamental semi-invariants to the subspace W_T. As they are monomial, also the restriction of the canonical equation is a monomial. □

8.6 The serie $B_{n,l}$

In this subsection we introduce a serie of quivers, which covers Dynkin diagrams of type B_n. Moreover we give a real Schur root for each member of the Serie, hence they give rise to a reductive prehomogeneous vector space, whose discriminant is given a divisor D by theorem 7.2.20. Furthermore we will determine the spectrum and the roots of the Bernstein-Sato polynomial.

When we change the direction of certain arrows, the serie will also cover the Dynkin diagrams C_n and the Star quiver $*_n$,which we will discuss in detail in chapter 9. The ideas, and arguments are independent of the chosen orientation, but for simplicity we discuss only one fixed orientation and leave the adjustments to the reader.

Definition 8.6.1. Let n, l be positive integers with $l \geq 2$ and $n \geq 3$. The quiver $B_{n,l}$ consists of n nodes, which we denote by $1, \ldots, n$, and $n-1$ arrows α_i with $t\alpha_i = i$ and $h\alpha_i = i+1$ for $i = 1, \ldots n-2$ and l arrows β_j with $t\beta_j = n-1$ and $h\beta_j = n$ for $j = 1, \ldots, l$
The case $l = 2$ for the quiver $B_{n,2}$ is shown in the following figure:

$$1 \xrightarrow{\alpha_1} 2 \xrightarrow{\alpha_2} \cdots \xrightarrow{\alpha_{n-2}} n-1 \underset{\beta_2}{\overset{\beta_1}{\rightrightarrows}} n$$

We simplify our notation and write the symbol $\overset{\beta_l}{\rightrightarrows}$ for the l arrows $\beta_j : n-1 \to n$. When considering a representation M, we will use the capital letter A_i (resp. B_i) for the linear maps attached to the arrows α_i (resp. β_i), i.e. a representation M with dimension vector $\mathbf{d} \in \mathbb{N}^n$ is written as

$$\mathbb{C}^{\mathbf{d}_1} \overset{A_1}{\to} \mathbb{C}^{\mathbf{d}_2} \overset{A_2}{\to} \cdots \overset{A_{n-2}}{\to} \mathbb{C}^{\mathbf{d}_{n-1}} \overset{B_l}{\rightrightarrows} \mathbb{C}^{\mathbf{d}_n}.$$

Lemma 8.6.2. For $\mathbf{d} = (l, \ldots, l, 1)$ we have:

1. $q_{B_{n,l}}(\mathbf{d}) = 1$.

2. $\mathbf{d}$ is a real Schur root.

In particular $(\mathrm{GL}(B_{n,l}), \mathbf{d}, \mathrm{Rep}(B_{n,l}, \mathbf{d})$ is a reductive prehomogeneous vector space.

Proof. The first statement follows by a direct calculation. Indeed we have

$$\begin{aligned} q_{b_{n,l}}(\mathbf{d}) &= \sum_i \mathbf{d}_i - \sum_{\alpha_i} \mathbf{d}_{t\alpha_i}\mathbf{d}_{h\alpha_i} - \sum_{\beta_i} \mathbf{d}_{t\beta_i}\mathbf{d}_{h\beta_i} \\ &= [(n-1)l^2 + 1] - \left[\sum_{\alpha_i} l^2 + \sum_{\beta_i} l\right] \\ &= (n-1)l^2 + 1 - (n-2)l^2 - l^2 = 1. \end{aligned}$$

Before we prove the second statement, we recall that $\mathbf{d}$ is a real Schur root if and only if there is a representation $M \in \operatorname{Rep}(B_{n,l}, \mathbf{d})$

$$\mathbb{C}^l \xrightarrow{A_1} \mathbb{C}^l \xrightarrow{A_2} \dots \xrightarrow{A_{n-2}} \mathbb{C}^l \overset{B_l}{\rightrightarrows} \mathbb{C},$$

which is a brick. And by lemma 7.2.9 M is a brick if and only if

$$\operatorname{GL}(B_{n,l}), \mathbf{d})_M \cong \mathbb{C}^*.$$

Moreover we remark that for any given representation M an element

$$g = (G_1, \dots, G_{n-1}, \lambda) \in \operatorname{GL}(B_{n,l}, \mathbf{d}) = \prod_{i=1}^{n-1} \operatorname{GL}_l(\mathbb{C}) \times \operatorname{GL}_1(\mathbb{C})$$

is by definition an element of the isotropy group $\operatorname{GL}(B_{n,l}, \mathbf{d})_M$ if and only if the following equations are satisfied:

$$G_{i+1} A_i G_i^{-1} = A_i, \text{ for } i = 1, \dots, n-2 \quad (8.2)$$
$$\lambda B_j G_{n-1}^{-1} = B_j, \text{ for } j = 1, \dots, l. \quad (8.3)$$

Now to prove the statement we consider the following representation M: The linear maps A_i are given by the identity matrix $\mathbf{1}_l \colon \mathbb{C}^l \to \mathbb{C}^l$, for $i = 1, \dots, n-2$, and the linear map $B_j \colon \mathbb{C}^l \to \mathbb{C}$ is given by the matrix $(0, \dots, 1, \dots, 0) = e_j^T$, i.e. the transpose of the j-th standard basis vector of $\mathbb{C}^l$, for $j = 1, \dots, l$.
We claim that M is a brick and hence $\mathbf{d}$ is a real Schur root.
Let $g = (G_1, \dots, G_{n-1}, \lambda) \in \operatorname{GL}(B_{n,l}, \mathbf{d})_M$. Then we conclude from the equations (8.3) that $G_{n-1} = \lambda \mathbf{1}_l$ and from the equations (8.2),that $G_i = G_{i+1}$ for $i = 1, \dots, n-2$. Hence we have $g = (\lambda \mathbf{1}_l, \dots \lambda \mathbf{1}_l, \lambda)$, i.e. $\operatorname{GL}(B_{n,l}, \mathbf{d}) \cong \mathbb{C}^*$, so $\mathbf{d}$ is a real Schur root. □

The dimension vector $\mathbf{d} = (l, \dots, l, 1)$ is in a unique with the property $q_Q(\mathbf{d}) = 1$, if we fix the dimensions for the nodes n and $n-1$ with $\mathbf{d}_{n-1} = l$ and $\mathbf{d}_n = 1$.

Lemma 8.6.3. Let $n, m, l > 1$. A dimension vector $\mathbf{d} = (m, l, \dots, l, 1)$ for the quiver $B_{n+1,l}$ fulfils $q_Q(\mathbf{d}) = 1$ if and only if $m = l$.

Proof. We compute

$$\begin{aligned} q_Q(\mathbf{d}) &= \sum_{i \in Q_0} \mathbf{d}_i^2 - \sum_{\alpha \in Q_1} \\ &= (m^2 + n * l^2 + 1^2) - (l \cdot m + (n-1) \cdot l^2 + l \cdot l) \\ &= 1 + m^2 - l \cdot m \end{aligned}$$

and conclude that $q_Q(\mathbf{d}) = 1$ if and only if $l = m$. □

Remark 8.6.4. If we do not fix the dimensions at the node $n-1$, then there exists dimension vectors $\mathbf{d}$, which fulfil the condition $q_Q(\mathbf{d}) = 1$. E.g. the dimension vector $\mathbf{d} = (2,1,1)$ for the quiver

$$\bullet \to \bullet \rightrightarrows \bullet$$

satisfies this condition. However in this example every representation

$$\mathbf{C}^2 \xrightarrow{A} \mathbf{C}^1 \overset{B_1,B_2}{\rightrightarrows} \mathbf{C}^1$$

decomposes and hence does not lead to a prehomogeneous vector space.

Remark 8.6.5. Again we use the following convention, when dealing with functions on the representation space. We identify $\operatorname{Rep}(Q,\mathbf{d})$ with $\mathbf{C}^n$ by the basis given by the elementary matrices, ordered lexicographically, and consider expressions like $\det(A)$ as a function on $\operatorname{Rep}(Q,\mathbf{d})$, where each entry of the matrix is a coordinate function. E.g. in the situation of $B_{3,2}$ the expression $\det(A)$ would stand for the following vanishing set

$$\det(A) = \{(A, B_1, B_2) \in \operatorname{Rep}(B_{3,2}, \mathbf{d}) \mid \det(A) = 0\}.$$

Proposition 8.6.6. The fundamental semi-invariants of the reductive prehomogeneous vector space $(\operatorname{GL}(B_{n,l}, \mathbf{d}, \operatorname{Rep}(B_{n,l}, \mathbf{d}))$ are given by the the equations

$$\begin{aligned} h_i &= \det(A_i),\ i = 1, \ldots, n-2 \\ h_{n-1} &= \det(B), \end{aligned}$$

where $B = (B_1|\ldots|B_l)$ is the concatenation of the matrices B_j, $j = 1,\ldots l$. In particular they are defining equations for the irreducible components of the discriminant.

Proof. By Kac's result 7.3.8 there are precisely $n-1$ fundamental semi-invariants. We conclude from lemma 7.4.3 that $n-2$ of them are given by the equations $h_i = \det(A_i)$ for $i = 1, \ldots, n-2$ and by lemma 7.4.6 the equation $h_{n-1} = \det(B)$ is $n-1$–th one. □

Let $X = (x_{ij})$ be the $l \times l$ matrix in the indeterminates x_{ij} and $\partial_X = (\frac{\partial}{\partial_{x_{ij}}})$ be the corresponding matrix of partial derivatives. Then the determinant $\det(\partial)$ is a differential operator and the classical *Cayley identity* states (see [CSS13]):

$$\det(\partial_X)(\det(X))^{s+1} = \prod_{i=1}^{l}(s+i)(\det(X))^s.$$

Hence Bernstein-Sato polynomial of the

Proposition 8.6.7. The Bernstein-Sato polynomial of the reduced discriminant h^{red} of $B_{n.l}$ is:

$$b_{h^{red}}(s) = \prod_{i=1}^{l} (s+i)^n.$$

Proof. By proposition the reduced discriminant is a product of $l \times l-$matrices $\det(A_1), \ldots, \det(A_{n-2}), \det(A_{n-1})$ (with $A_{n-1} := B$), whose entries are independent variables. Hence we have

$$\det(\partial_{A_i}) \det(A_j) = \det(A_j) \det(\partial_{A_i})$$

for $i \neq j$ and the statement is a direct consequence of the Cayley identity. □

Corollary 8.6.8. The spectrum at $t = 0$ is given by

$$\nu_{k \cdot n+j} = nl^2 - nkl - nl + kn + j - 1$$

for $k = 0, \ldots, l-1$ and $j = 1, \ldots, n$.

Remark 8.6.9. We remark that all statements hold, if we choose any other orientation for each arrow α_i or choose the opposite direction for all β_j, i.e. $h\beta_j = n - 1$ and $t\beta_j = n$ for $j = 1, \ldots, l$. In the case $h\beta_j = n - 1$ and $t\beta_j = n$ for $j = 1, \ldots, l$ the Dynkin diagram C_n is part of the serie. If we change the orientation of all arrows β_j only, then the matrix B in proposition 8.6.6 is given by the transposition of the old one.

Chapter 9

The spectrum of the star quiver

The *star quiver* $\star_n$ consists of $n+1$ sources Q_l and one sink Q. For each source Q_l there is one oriented arrow from the source to the sink $\overrightarrow{Q_l Q}$. We consider the dimension vector $\mathbf{d}$ with $\mathbf{d}(Q) = n$ and $\mathbf{d}(Q_l) = 1$. The star quiver $\star_5$ with $d = (5, 1, 1, 1, 1, 1, 1)$ is shown in the following figure

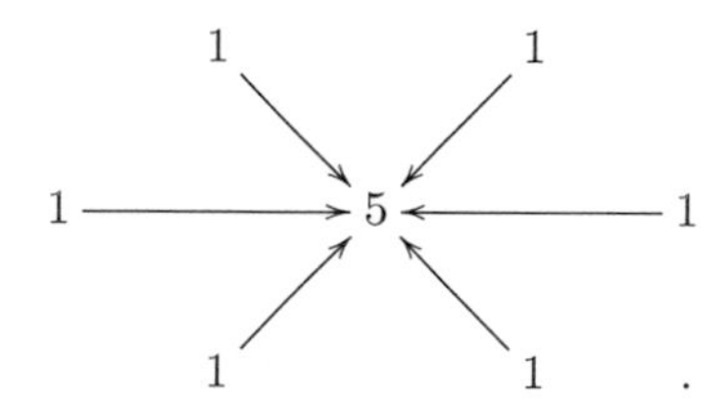

The representation space $V = \operatorname{Rep}(\star_n, \mathbf{d})$ of $\star_n$ can be identified with space of $n \times n+1$ matrices $(x_{ij})_{ij} \in \operatorname{Mat}_{n \times n+1}(\mathbb{C})$, where each column $(x_{ij})_l$ corresponds to a linear map $V(Q_l) \to V(Q)$.
By [BM06] proposition 7.4 the pair $(\star_n, \mathbf{d})$ gives rise to a linear free divisor, so in particular it defines a prehomogeneous vector space whose discriminant is given by a reduced equation. Precisely the group $\operatorname{GL}(\star_n, \mathbf{d}) = \operatorname{GL}_n(\mathbb{C}) \times \operatorname{GL}_1^{n+1}(\mathbb{C})$ acts on $\operatorname{Rep}(\star_n, \mathbf{d}) = \operatorname{Mat}_{n \times n+1}(\mathbb{C})$ by

$$(A, \lambda_1, \ldots, \lambda_{n+1}).M = AM\operatorname{diag}(\lambda_1, \ldots, \lambda_{n+1})^{-1}$$

and its discriminant D is given by the product of the determinants of all maximal minors of the $n \times (n+1)$ matrix $\Delta = (x_{ij})_{ij}$.

9.1 A solution to the Birkhoff problem

In this section we will construct a basis of the FL-Gauss-Manin system of the star quiver, which solves the Birkhoff problem, i.e. we are looking for a basis $\omega = (\omega_0, \ldots, \omega_{n(n+1)-1})$ of $G(*D)$, such that for $0 \leq i \leq n(n+1) - 1$ holds

$$\tau\partial_\tau\omega_i = -\tau\omega_{i+1} - \nu_i\omega_i,$$

where we have set $\omega_{n(n+1)} = t\omega_0$ and the ν_i and t are complex numbers. Let us assume for a moment that we knew the ν_i and the t.
We remark that a solution the Birkhoff problem is equivalent to the following action of the operators $(\tau\partial_\tau + \nu_i)$:

$$(\tau\partial_\tau + \nu_i)\omega_i = -\tau\omega_{i+1}.$$

If such a basis exists, then by lemma 3.2.2 and proposition 5.2.4 it can be writen as $\omega_i = b_i\alpha$ with $b_i \in \mathcal{O}_V$ and $\deg(b_i) = i$.[1] Hence we are looking for polynomials b_i satisfying

$$(\tau\partial_\tau + \nu_i)b_i\alpha = -\tau b_{i+1}\alpha.$$

In lemma 9.1.4 we are going to prove such a relation, where the polynomials b_i are given by determinants of minors of the matrix Δ and minors of a matrix A, which corresponds to the linear function f. Afterwards we will construct a basis solving the Birkhoff problem, but before we need to consider special vector fields.

We denote by $\Delta^{j_1\ldots j_n}_{1\ldots n}$ the determinant of the maximal minor of Δ, given by the columns $j_1, \ldots, j_n$ and by

$$r_i\partial_j := \sum_{k=1}^{n+1} x_i^k \partial_{x_j^k} := \sum_{k=1}^{n+1} x_{ik}\partial_{x_{jk}}$$

The vector fields

$$\begin{aligned} \eta_i &:= -r_1\partial_{r_1} - \ldots - r_{i-1}\partial_{i-1} + (n-1)r_i\partial_i - r_{i+1}\partial_{i+1} - \ldots - r_n\partial_n \\ \xi_{ij} &:= r_i\partial_{r_j},\ i \neq j. \end{aligned}$$

belong to $\mathrm{Der}(-log\ \Delta)_0$. Indeed as $r_i\partial_{r_j}$ acts on $\Delta^{j_1\ldots j_n}_{1\ldots n}$ by replacing the $j-$th row by the $i-$th row, we conclude

$$\begin{aligned} \eta_i(\Delta^{j_1\ldots j_n}_{1\ldots n}) &= -(n-1)\Delta^{j_1\ldots j_n}_{1\ldots n} + (n-1)\Delta^{j_1\ldots j_n}_{1\ldots n} = 0 \\ \xi_{ij}(\Delta^{j_1\ldots j_n}_{1\ldots n}) &= \Delta^{j_1\ldots j_n}_{1\ldots j\mapsto i\ldots n} = 0. \end{aligned}$$

[1] A pri ori the b_i lie in $\mathcal{O}_V[\tau]$, but as in the proof of proposition 4.5 (i) [GMS11], they can be chosen to lie in $\mathcal{O}_V$.

For later applications we also need to know, how these vector fields act on determinants of arbitrary $m \times m-$minors of Δ. Therefore we denote for two index sets $I = (i_1, \dots, i_m)$ of rows and $J = (j_1, \dots, j_m)$ of columns the corresponding determinant of the minor by

$$\Delta_I^J = \det(x_i^j) = \det(x_{ij})_{i \in I, j \in J}.$$

and a map by

$$\begin{aligned} \Phi_I(i) \colon I \quad &\to \quad \{1, \dots, m\} \\ i_\alpha \quad &\mapsto \alpha. \end{aligned}$$

Lemma 9.1.1. We have

$$\eta_i(\Delta_I^J) = \begin{cases} (n-m)\Delta_I^J & \text{if } i \in I \\ -m\Delta_I^J & \text{if } i \notin I \end{cases}$$

$$\xi_{ij}(\Delta_I^J) = \begin{cases} 0 & \text{if } j \notin I \\ (-1)^{1-\Phi_I(i)+\Phi_I(j)}\Delta_{(ij)I}^J & \text{if } j \in I, i \notin I \\ 0 & \text{if } j, i \in J, \end{cases}$$

where the index $(ij)I$ means : replace j by i in the ordered tuple I.

Proof. To determine $\eta_i(\Delta_I^J)$ we calculate

$$\begin{aligned} r_i \partial_{r_i}(\Delta_I^J) &= (\sum_{k=1}^{n+1} x_i^k \partial_{x_i^k})(\sum_{\sigma \in \mathrm{S}_m} \mathrm{sgn}(\sigma) \prod_{l=1}^{m} x_{i_l}^{j_{\sigma(l)}}) \\ &= \sum_{\sigma \in \mathrm{S}_m} \mathrm{sgn}(\sigma) \sum_{k=1}^{n+1} x_i^k \sum_{p=1}^{m} (\partial_{x_i^k} x_{i_p}^{j_{\sigma(p)}}) \prod_{l \neq p} x_{i_l}^{j_{\sigma(l)}} \\ &= \sum_{\sigma \in \mathrm{S}_m} \mathrm{sgn}(\sigma) \sum_{k=1}^{n+1} x_i^k \sum_{p=1}^{m} \delta_{i,i_p} \delta_{k,j(p)} \prod_{l \neq p} x_{i_l}^{j_{\sigma(l)}} \\ &= \begin{cases} \Delta_I^J, \text{ if } i \in I \\ 0, \text{ if } i \notin I. \end{cases} \end{aligned}$$

The statement for $\xi_{ij}(\Delta_I^J)$ is obtained in a similar way. □

Let $f = \sum_{i,j} a_j^i x_i^j$ be a linear function on V. Obviously f lies in the open orbit $V^* \backslash D^*$ if and only if the determinant of each maximal minors of $A = (a_{ij} = a_j^i)$ is not zero. In the following we will identify f by its matrix A. We write

$$f_i^j = \sum_{l=1}^{n+1} a_l^j x_i^l.$$

Lemma 9.1.2. We have the following identities

$$f_i^j = \begin{cases} \frac{1}{n}(f + \eta_i(f)), & \text{if } i = j \\ \xi_{ij}(f), & \text{if } i \neq j. \end{cases}$$

We need to introduce additional notations. For an $m-$tuple $\tilde{I} = (i_1, \ldots, i_m)$ we define

$$\begin{aligned} \tilde{I} + i &:= (i_1, \ldots, i_m, i) \\ \tilde{I} - i_j &:= (i_1, \ldots, \widehat{i_j}, \ldots, i_m). \end{aligned}$$

For two index sets $\tilde{I}$ of rows and $\tilde{J}$ of columns with the same number of elements, we denote by $\Delta_{\tilde{I}}^{\tilde{J}}$ (resp. $A_{\tilde{I}}^{\tilde{J}}$) the determinant of the matrix given by the entries of Δ (resp. A) indexed by $\tilde{I}$ and $\tilde{J}$.

Let $K \subset \{1, \ldots, n\}$ and $L \subset \{1, \ldots, n+1\}$ subsets of $n+1-m$ distinct elements and $I \subset \{1, \ldots, n\}$ a subset with $m+1$ elements. We denote by $J = \{1, \ldots, n+1\} - L = C(L)$ the complement of L.

Remark 9.1.3. As $|K| = n-m+1$ and $|I| = m+1$, we have $|K|+|I| = n+2$ and because I and K are subsets of $\{1, \ldots, n\}$, we have

$$|K \cap I| \geq 2.$$

Lemma 9.1.4. If the determinant A_L^K does not vanish, then we have in $G(*D)$ for any $l \in L$:

$$(\tau\partial_\tau + (n+1)m)\Big[\frac{1}{n}\sum_{k \in K \cap I}(-1)^{m+1+\Phi_I(k)} A_{L-l}^{K-k}\Delta_{I-k}^{J}\Big]\alpha = -\tau(A_L^K \Delta_I^{J+l})\alpha$$

where $\alpha = \iota_E(vol)$.

Remark 9.1.5. Because we are investigating properties of the coefficients of the generator α only, we will drop for simplicity the generator α in most of the following formulas.

Proof. Let $I = (i_1, \ldots, i_{m+1})$ and $J = (j_1, \ldots, j_m)$. For $k \in K$ we consider the determinant

$$\Gamma_I^{J+k} = \begin{vmatrix} x_{i_1}^{j_1} & \cdots & x_{i_1}^{j_m} & f_{i_1}^k \\ \vdots & & \vdots & \vdots \\ x_{i_{m+1}}^{j_1} & \cdots & x_{i_{m+1}}^{j_m} & f_{m+1}^k \end{vmatrix}.$$

We recall the definition $f_i^j = \sum_{l=1}^{n+1} a_l^j x_i^l$ and conclude from the assumption $L = C(J)$

$$\Gamma_I^{J+k} = \sum_{l \in L} a_l^k \Delta_I^{J+l}.$$

Using Cramer's rule we obtain the formula

$$A_L^K \Delta_I^{J+l} = \sum_{k \in K} A_{L-l}^{K-k} \Gamma_I^{J+k}. \tag{9.1}$$

We expand Γ_I^{J+k} in the last column and obtain

$$\Gamma_I^{J+k} = \sum_{i \in I} (-1)^{m+1+\Phi_I(i)} \Delta_{I-i}^J f_i^k. \tag{9.2}$$

We combine the equations and apply lemma (9.1.2) to get

$$\begin{aligned}
A_L^K \Delta_I^{J+l} &\overset{(9.1)}{=} \sum_{k \in K} A_{L-l}^{K-k} \Gamma_I^{J+k} \\
&\overset{(9.2)}{=} \sum_{k \in K} A_{L-l}^{K-k} \sum_{i \in I} (-1)^{m+1+\Phi_I(i)} f_i^k \Delta_{I-i}^J \\
&= \sum_{k \in K} \sum_{i \in I-k} (-1)^{m+1+\Phi_I(i)} A_{L-l}^{K-k} \Delta_{I-i}^J f_i^k \\
&+ \sum_{k \in I \cap K} (-1)^{m+1+\Phi_I(i)} A_{L-l}^{K-k} \Delta_{I-k}^J f_k^k \\
&\overset{(9.1.2)}{=} \sum_{k \in K} \sum_{i \in I-k} (-1)^{m+1+\Phi_I(i)} A_{L-l}^{K-k} \Delta_{I-i}^J \xi_{ik}(f) \\
&+ \sum_{k \in K \cap I} (-1)^{m+1+\Phi_I(k)} A_{L-l}^{K-k} \Delta_{i-k}^J \frac{1}{n} (f + \eta_k(f)).
\end{aligned}$$

Now we multiply the last equation with τ, use from the Fourier Laplace transformation the fact $-\partial_\tau = r = f$ and apply the division lemma (lemma 5.2.3), i.e. in $G(*D)$ we have

$$\tau g \xi(f) \alpha = \xi(g) \alpha$$

for all $\xi \in \mathfrak{a} = \mathrm{Der}(-log\ \Delta)_0$ and $g \in \mathcal{O}_V$. Hence we get

$$\begin{aligned}
\tau A_L^K \Delta_I^{J+l} &= \sum_{k \in K} \sum_{i \in I-k} (-1)^{m+1+\Phi_i(i)} A_{L-l}^{K-k} \xi_{ik}(\Delta_{I-i}^J) \\
&- \tau \partial_\tau \frac{1}{n} \sum_{k \in K \cap I} (-1)^{m+1+\Phi_I(k)} A_{L-l}^{K-k} \Delta_{I-k}^J \\
&+ \frac{1}{n} \sum_{k \in K \cap I} (-1)^{m+1+\Phi_I(k)} A_{L-l}^{K-k} \eta_k(\Delta_{I-k}^J).
\end{aligned}$$

We apply lemma 9.1.1 and obtain

$$\begin{aligned}\tau A_L^K \Delta_I^{J+l} &= -\tau\partial_\tau\Big[\frac{1}{n}\sum_{k\in K\cap I}(-1)^{m+1+\Phi_I(k)}A_{L-l}^{K-k}\Delta_{I-k}^J\Big] \\ &- \sum_{k\in K\cap I}(m+\frac{m}{n})(-1)^{m+1+\Phi_I(k)}A_{L-l}^{K-k}\Delta_{I-k}^J, \quad (9.3)\end{aligned}$$

which finishes the proof. □

We remark that the factors $(-1)^{m+1+\Phi_I(k)}$ in the equations are determined by the datum Δ_{I-k}^J, as $|J| = m$. Hence to simplify the notation we set by abuse of notation

$$\Delta_{I-k}^J := (-1)^{m+1+\Phi_I(k)}\Delta_{I-k}^J.$$

Moreover we will see that the sign will have no effect concerning the spectrum.
We will now describe a construction of a basis $\underline{\omega} = (\omega_0, \dots, \omega_{n(n+1)-1})$ solving the Birkhoff problem for the star quiver.
First we will explain, how to construct the first n coefficients b_i, for $i = 1, \dots, n$, i.e. we construct the first n elements $\omega_0, \dots, \omega_{n-1}$ of the basis $\underline{\omega}$.
We start with a maximal minor $b_n = \Delta_{I_1}^{J_1}$ of Δ and use the key-lemma 9.1.4 to construct b_{n-1}. Precisely, let

$$\begin{aligned} I_1 &= (1, \dots, n) \\ J_1 &= (j_1, \dots, j_n) = (1, \dots, \widehat{j_{n+1}}, \dots, n+1).\end{aligned}$$

We put $L_1 = (j_{n+1}, j_n)$ and choose $K_1 = (k_1, k_2)$, such that $A_{L_1}^{K_1} \neq 0$. Because K_1 is a pair of two distinct elements of $\{1, \dots, n\}$ we have $K_1 \cap I_1 = \{k_1, k_2\}$. We conclude from lemma 9.1.4

$$-n\tau A_{L_1}^{K_1}\Delta_{I_1}^{J_1} = (\tau\partial_\tau + (n+1)(n-1))\Big[A_{L_1-j_n}^{K_1-k_1}\Delta_{I_1-k1}^{J_1-j_n} + A_{L_1-j_n}^{K_1-k_2}\Delta_{I_1-k_2}^{J_1-j_n}\Big].$$

Hence if we put $b_{n-1} = B_1 + B_2 := \frac{1}{n}\Big[A_{L_1-j_n}^{K_1-k_1}\Delta_{I_1-k1}^{J_1-j_n} + A_{L_1-j_n}^{K_1-k_2}\Delta_{I_1-k_2}^{J_1-j_n}\Big]$, we have

$$(\tau\partial_\tau + (n+1)(n-1))b_{n-1}\alpha = -n\tau b_n\alpha,$$

i.e.

$$(\tau\partial_\tau + (n+1)(n-1))\omega_{n-1} = -n\tau\omega_n.$$

We proceed by applying to each summand $B_i = \frac{1}{n}A_{L_1-j_n}^{K_1-k_i}\Delta_{I_i-k_i}^{J_1-j_n}$ the lemma 9.1.4. In the $r-$th step, for $r \geq 2$, we set

$$\begin{aligned} L_r &:= (j_{n+1}, \dots, j_{n+1-r}) \\ l_r &:= j_{n+1-r}\end{aligned}$$

and choose

$$K_r = (k_1, \ldots, k_{r+1})$$

such that $A_{L_r}^{K_r} \neq 0$. Moreover we set for $1 \leq s \leq r$

$$\begin{aligned} I_r^s &= (1, \ldots, n) - (k_1, \ldots, \widehat{k_s}, \ldots, k_r), \\ J_r &= (j_1, \ldots, j_{n+1-r}), \\ K_r^s &= (k_1, \ldots, k_r) - k_s \end{aligned}$$

and remark that $K_r \cap I_r^s = \{k_s, k_{r+1}\}$. Hence we can reformulate lemma 9.1.4 as :

Corollary 9.1.6. With the assumptions and notation[2] from above we have

$$\begin{aligned} -n\tau A_{L_r}^{K_r} \Delta_{I_r^s}^{J_r} &= \left[\tau\partial_\tau + (n+1)(n-r)\right]\left[A_{L_r-l_r}^{K_r-k_{r+1}} \Delta_{I_r^s-k_{r+1}}^{J_r-l_r} + A_{L_r-l_r}^{K_r-k_s} \Delta_{I_r^s-k_s}^{J_r-l_r}\right] \\ &= \left[\tau\partial_\tau + (n+1)(n-r)\right]\left[A_{L_{r+1}}^{K_{r+1}^{r+1}} \Delta_{I_{r+1}^s}^{J_{r+1}} + A_{L_{r+1}}^{K_{r+1}^s} \Delta_{I_{r+1}^{r+1}}^{J_{r+1}}\right]. \end{aligned} \tag{9.4}$$

We simplify the notation and write

$$\omega_{n-r+1} = b_{n-r+1}\alpha = \sum B_i \alpha,$$

where the B_i of the $r-$th step are of the form $cA_{\tilde{L}_i}^{\tilde{K}_i} \Delta_{\tilde{I}_i}^{\tilde{J}_i}$. Here c is a power of $\frac{1}{n}$. By applying corollary 9.1.6 to each B_i we get (in the $r-$th step) a family of pairs (B_i^1, B_i^2), which fulfils

$$-\tau B_i = (\tau\partial_\tau + (n+1)(n-r))(\frac{1}{n}B_i^1 + \frac{1}{n}B_i^2).$$

Hence by putting $b_{n-(r+1)} := \frac{1}{n}\sum B_i^1 + B_i^2$ we get

$$-\tau\omega_{n-r} = (\tau\partial_\tau + (n+1)(n-r))\omega_{n-(r+1)}$$

and the connection ∇_{∂_τ} acts on the first n elements of the basis $\omega_1, \ldots \omega_{n-1}$ in the desired way. We extend these elements to a basis as follows:

After we have done the calculations from above for all irreducible components Δ_l, we have a family $b_i^{(l)}$ with $b_n^{(l)} = \Delta_l$, such that

$$(\tau\partial_\tau + (n+1)i)b_i^{(l)} = -n\tau b_{i+1}^{(l)}$$

holds for $1 \leq l \leq n+1$ and $1 \leq i < n$. Moreover we have $\deg(b_i^{(l)}) = i$. We put for $1 \leq l \leq n+1$

$$\omega_{ln} = \prod_{j=1}^{l} \Delta_j \alpha$$

[2]We recall our simplification dropping α in the equations.

and for $1 \leq i \leq n-1$ and $1 \leq l \leq n$

$$\omega_{ln+i} = b_i^{(l+1)} \prod_{j=1}^{l} \Delta_j \alpha.$$

Furthermore we exchange $\omega_{n(n+1)} = \prod_{j=1}^{l+1} \Delta_j \alpha$ by $\omega_0 = 1$.

Proposition 9.1.7. The basis $\underline{\omega}$ from above is a solution to the Birkhoff problem, i.e.

$$-\tau \partial_\tau(\omega_i) = \tau \omega_{i+1} + \nu_i \omega_i,$$

where $\nu_i \in \mathbb{C}$.

Proof. That $\underline{\omega}$ is a basis, follows from the discussion in the previous sections. From the proof of proposition 5.2.4 we knew that a basis of $G(*D)$ and $G_0(log\ D)$ can be chosen from $\frac{\Omega^{n-1}_{V/T}(log\ D)}{df \wedge \Omega^{n-2}(log\ D)}$. By proposition 3.2.6 this module is isomorphic to $\frac{\mathcal{O}_V}{df(\mathrm{Der}(-log\ h)}\alpha$ and by the proof of proposition 4.1.11 it is generated $\underline{\omega}$.
Hence it remains to prove that $\underline{\omega}$ solves the Birkhoff problem. We will prove this in three steps, but before we recall:

- The operator ∂_τ acts as multiplication by $-f$ on $G(*D)$.
- The division lemma (lemma 5.2.3): In $G(*D)$ holds

  $$g\xi(f)\alpha = \tau g \xi(f)$$

  for all $\xi \in \mathrm{Der}(-log\ \Delta)_0$ and $g \in \mathcal{O}_V$.

1. First we prove that $\tau\partial_\tau(\omega_0) = \tau\omega_1 + \nu_0\omega_0$ holds.
   Therefore we write

   $$f = b_1^{(1)} + 1 \cdot \xi_0(f),$$

   with $\xi_0 \in \mathrm{Der}(-log\ \Delta)_0$. Hence in $G(*D)$ we have

   $$\tau\partial_\tau(\omega_0) = -\tau f \alpha = -\tau b_1^{(1)}\alpha - \tau\xi_0(f)\alpha.$$

   We apply the division lemma and conclude

   $$\begin{aligned} \tau\partial_\tau(\omega_0) &= -\tau b_1^{(1)}\alpha - \xi_0(1)\alpha \\ &= -b_1^{(1)}\alpha = \tau\omega_1. \end{aligned}$$

2. Next we prove that $\tau\partial_\tau\omega_{ln} = -\tau\omega_{ln+1} - \nu_{nl}\omega_{ln}$ holds for $1 \leq l \leq n$. Therefore we write as above

$$f = b_1^{(l+1)} + \xi_l(f)$$

with $\xi_l \in \mathrm{Der}(-log\ \Delta)_0$. We recall that the irreducible components Δ_l are semi-invariants, i.e. for any $\xi \in \mathrm{Der}(-log\ \Delta)$ we have[3]

$$\xi(\Delta_i) = d\chi_{\Delta_i}(\xi)\Delta_i,$$

where $d\chi_{\Delta_i}$ is the differential of the character corresponding to Δ_i. Hence we have by the Leibniz rule:

$$\xi(\prod_{j=1}^{l}\Delta_j) = (\sum_{j=1}^{l} d\chi_{\Delta_j}(\xi))\prod_{j=1}^{l}\Delta_j.$$

Using these equations and the division lemma we get

$$\begin{aligned}
\tau\partial_\tau(\omega_{ln}) &= -\tau f\prod_{j=1}^{l}\Delta_j\alpha \\
&= -\tau(b_1^{(l+1)} + \xi_l(f))\prod_{j=1}^{l}\Delta_j\alpha \\
&= -\tau b_1^{(l+1)}\prod_{j=1}^{l}\Delta_j\alpha - \xi_l(\prod_{j=1}^{l}\Delta_j)\alpha \\
&= -\tau\omega_{ln+1} - \lambda_l\omega_{ln},
\end{aligned}$$

where $\lambda_l = \sum_{j=1}^{l} d\chi_{\Delta_j}(\xi_l)$.

3. The remaining part will follow directly from corollary 9.1.6. For $1 \leq i \leq n-1$ and $1 \leq l \leq n$ we have

$$\begin{aligned}
\tau\partial_\tau\omega_{ln+i} &= \tau\partial_\tau b_i^{(l)}\prod_{j=1}^{l}\Delta_j\alpha \\
&= -(\tau b_{i+1}^{(l)} + (n+1)i)\prod_{j=1}^{l}\Delta_j\alpha \\
&= -\tau\omega_{ln+i+1} - (n+1)i\omega_{ln+i}
\end{aligned}$$

[3]Here some additional explanation might be useful. Because D is a linear free divisor, we have an isomorphism $\mathrm{Der}(-log\ \Delta) \cong \mathcal{O}_V \otimes \mathfrak{g}$. We extend the differentials of the characters $\mathcal{O}_V$−linear and understand the action of a logarithmic vector field on a semi-invariant by this isomorphism.

□

We conclude from the proof above.

Corollary 9.1.8. In the basis $\underline{\omega}$ the connection is given by the matrix

$$\tau\partial_\tau\underline{\omega} = \underline{\omega}(\tau A_0(t) + A_\infty)d\tau,$$

where the matrices are

$$A_0 = \begin{pmatrix} 0 & 0 & \dots & 0 & t \\ 1 & 0 & \dots & 0 & 0 \\ 0 & 1 & \dots & 0 & 0 \\ \dots & \dots & \dots & \dots & \\ 0 & 0 & \dots & 1 & 0 \end{pmatrix},$$

$$A_\infty = \operatorname{diag}\begin{pmatrix} 0, & (n+1), & 2(n+1), & \dots & (n-1)(n+1), \\ \lambda_1, & (n+1), & 2(n+1), & \dots & (n-1)(n+1), \\ \lambda_2, & (n+1), & 2(n+1), & \dots & (n-1)(n+1), \\ \vdots & \vdots & \vdots & \dots & \vdots \\ \lambda_n, & (n+1), & 2(n+1), & \dots & (n-1)(n+1) \end{pmatrix}$$

and the lambdas are given as follows:

1. Determine $\xi_l \in \mathrm{Der}(-log\ h)_0$ and $g_l \in \mathbb{C}[V] = \mathbb{C}[\mathrm{Rep}(*_n, \mathbf{d})]$, such that
$$-f = b_1^{(l)} + g_l\xi_l(f)$$
holds.
2. Then we have $\lambda_l = g_l\xi_l(\prod_{j=1}^{l-1} h_j)$, where $h_j = \Delta_{(1,\dots,n)}^{(1,\dots,\hat{j},\dots,n+1)}$ are the equations of the irreducible components of the discriminant.

Remark 9.1.9. To compute the lambdas, we need to consider a different basis. Precisely we will put $\omega_n = h_{n+1} = \Delta_{(1,\dots,n)}^{(1,\dots,n)}$ and continue with $\omega_{2n} = h_{n+1}h_1$, $\omega_{3n} = h_{n+1}h_1h_2, \dots$. We remark that this does not affect the arguments of this section and the statements remain true.

9.2 Specializing the linear function

In this section we want to determine the lambdas from the previous section. Therefore we consider the case, when linear function f given by the matrix

$$A = \begin{pmatrix} 1 & 0 & \dots & 0 \\ 0 & 1 & \dots & 0 \\ \dots & \dots & \dots & \dots \\ 0 & 0 & \dots & 1 \\ 1 & 1 & \dots & 1 \end{pmatrix}.$$

By [Sev11] Lemma 6 all other cases (with $f \in V^\vee \backslash D^\vee$) are equivalent to this one.

For this choice of A corollary 9.1.6 becomes more simple for all irreducible components except the last one $h_{n+1} = \Delta^{(1,\dots,n)}_{(1,\dots,n)}$. This allows us by considering special vector fields to compute the lambdas from the previous section.

In the case that f is given by the matrix A, we can distinguish between two types of top minors: Either they contain the last column or not.

First we consider the case of a maximal minor that contains the last c are of the form, say $\Delta^{(1,\dots,\hat{j},\dots,n+1)}_{(1,\dots,n)}$ for $1 \leq j < n+1$. As $j_{n+1} = j \neq n+1$, we can choose $l_1 = j_n$, such that $K_1 = L_1$ and $A^{K_1}_{L_1}$ is the identity.

For example we can choose $j_n = j+1$, if $j \neq n$, and $j_n = j-1$ if $j = n$. We assume we had chosen j_n as one of this, then one of the two determinants $A^{K_1^i}_{L_1 - l_i}$ equals zero and the other one equals one. Hence we have

$$-n\tau A^{K_1}_{L_1} \Delta^{J_1}_{I_1} = (\tau\partial_\tau + (n+1)(n-1))(\Delta^{J_1 - k}_{I-k}),$$

where $k = j-1$ or $j+1$, but never $k = n+1$. In other words the tree defining the polynomials $b^{(l)}_{n-i}$ has in the first step only one branch.

We now apply the same principle in the next step, where we can choose l_2, such that $A^{K_2}_{L_2}$ is the identity[4] and again we have only one non vanishing subminor $\Delta^{J_2 - l_2}_{I_2 - l_2}$.

By induction we see that for the first class the tree is very simple, as it has only one branch in each step and it descends to a single variable.

The second class consisting of the minor $\Delta^{(1,\dots,n)}_{(1,\dots,n)}$ produces a tree with two branches in every step and descends to a certain sum of variables. However we can put it first in the basis and avoid this and further difficulties.

Let us do this calculation for an concrete example to illustrate the discussion from above.

1. We consider the first irreducible component $h_1 := \Delta^{(2,\dots,n+1)}_{(1,\dots,n)}$ and the linear function f given by the matrix A, i.e. we start with the index sets

[4] The rule is to avoid in every step $l_r = n+1$ and take $K_r = L_r$.

$$\begin{aligned} J_1 &= (2,\dots,n+1) \\ I_1 &= (1,\dots,n). \end{aligned}$$

Next we have to determine the index sets K_1 and L_1, such that the determinant $A_{L_1}^{K_1}$ does not vanish and we can apply corollary 9.1.6. We already knew $j_{n+1} = 1 \in L_1 = (j_{n+1}, j_n)$. As we want $A_{L_1}^{K_1}$ to be the determinant of the identity, we apply the rules from above, i.e. (1) $l_1 := j_n \neq n+1$ (respk. $l_i = i+1$) and (2) $K_1 = L_1$ (respk. $K_i = L_i$). We take $l_1 = j_n = 2$, so $K_1 = L_1 = (1,2)$ and obtain

$$A_{L_1-2}^{K_1^1} = A_{(1)}^{(1,2)-1} = 0$$

and

$$A_{L_1-2}^{K_1^2} = A_{(1)}^{(1)} = 1.$$

Hence $\Delta_{I_1}^{J_1}$ descends to $\Delta_{I_2}^{J_2}$.

2. In the next step we have the following index sets :

$$\begin{aligned} J_2 &= (3,\dots,n+1) \\ I_2 &= (1,3,\dots,n). \end{aligned}$$

Again we have to determine index sets K_2, L_2, such that $A_{L_2}^{K_2}$ is again the determinant of the identity. We knew that $1,2 \in L_2$ and as we only have to avoid by rule (1) $l_2 = n+1$, we set $l_2 = 3$. Hence we get by rule (2) :

$$\begin{aligned} L_2 &= (1,2,3) \\ K_2 &= (1,2,3). \end{aligned}$$

We have $I_2 \cap K_2 = \{1,3\}$ and by corollary 9.1.6 :

$$A_{(1,2,3)}^{(1,2,3)} \Delta_{I_2}^{J_2} \mapsto (A_{L_2-3}^{K_2^1} \Delta_{(1,\hat{2},3,\dots,n)-1}^{(3,\dots,n+1)-3}, A_{L_2-3}^{K_2^2} \Delta_{(1,\hat{2},3,\dots,n)-3}^{(3,\dots,n+1)-3}),$$

where

$$A_{L_2-3}^{K_2^1} = A_{(1,2,3)-3}^{(1,2,3)-1} = \begin{vmatrix} 0 & 0 \\ 0 & 1 \end{vmatrix} = 0$$

and

$$A_{L_2-3}^{K_2^2} = A_{(1,2)}^{(1,2)} = \left|\mathbf{1}\right| = 1.$$

Hence $\Delta_{I_2}^{J_2}$ descends to $\Delta_{I_3}^{J_3}$ with $J_3 = (4,\dots,n+1)$ and $I_3 = (1,4,\dots,n)$.

3. It is clear, how this procedure continues and that $\Delta_{I_1}^{J_1}$ descends to $\Delta_{(1)}^{(n+1)}$. We summarize this observation:

Lemma 9.2.1. $\Delta_{(1,\dots,n)}^{(1,\dots,\hat{j},\dots,n+1)}$ descends to $\Delta_{(j)}^{(n+1)} = x_j^{n+1}$ for every $1 \leq j < n+1$.

We recall that the basis constructed in the previous section the lambdas are given by corollary 9.1.8 as follows:

1. We determine $\xi_l \in \mathrm{Der}(-log\ h)_0$ and $g \in \mathbb{C}[V]$, such that we have
$$-f = b_1^{(l)} + g_l\xi_l(f).$$
2. Then we have $\lambda_l = g_l\xi_l(\prod_{j=1}^{l-1} h_j)$.

By lemma 9.2.1 we know the coefficients $b_1^{(l)} = x_l^{n+1}$ for $l \neq n+1$.
Because b_1^{n+1} is more complicated, we avoid solving $-f = b_1^{(n+1)} + g_{n+1}\xi_{n+1}(f)$ by reordering the basis. We start with $\omega_n = h_{n+1}$ instead of h_1 and proceed as $\omega_{2n} = h_{n+1}h_1,\ \omega_{3n}h_{n+1}h_1h_2, \dots$.
In the first step we need to find for each $b_1^{(l)} = \Delta_l^{(n+1)} = x_l^{n+1}$ a vector field ξ_l, such that
$$-f = \Delta_{(l)}^{(n+1)} + g_l\xi_l(f).$$
Therefore we consider the vector fields
$$\begin{aligned} c_i\partial_{c_i} &= \textstyle\sum_{k=1}^{n} x_k^i \partial_{x_k^i} \\ r_i\partial_{r_i} &= \textstyle\sum_{k=1}^{n+1} x_i^k \partial_{x_i^k} \\ \eta_j &= -\textstyle\sum_{k=1}^{n} r_k\partial_{r_k} + nr_j\partial_{r_j} \\ \xi_{ij} &= c_i\partial_{c_i} - c_j\partial_{c_j}. \end{aligned}$$
and obtain by a direct calculation:

Lemma 9.2.2. For $f = \sum_{k=1}^{n}(x_k^k + x_k^{n+1})$, $J_l = (1,\dots,\hat{l},\dots,n+1)$, $l \neq n+1$ and $I_1 = (1,\dots,n)$ we have the following formulas:

1. $f = n(n+1)\Delta_j^{(n+1)} - (n+1)\eta_j(f) + n\sum_{i=1}^{n+1}\xi_{ji}(f)$,
2. $\eta_j(\Delta_{I_1}^{J_1}) = 0$ and
3. $\xi_{ji}(\Delta_{I_1}^{J_l}) = \begin{cases} 0, & \text{if } i,j \neq l \\ \Delta_{I_1}^{J_l}, & \text{if } j \neq l, i \leq l \end{cases}$

Corollary 9.2.3. Let $j \neq l$, then for $\theta_j = -(n+1)\eta_j + n\sum_{i=1}^{n+1}\xi_{ji}$ we have :

$$\theta_j(\Delta^{(1,\dots,\hat{l},\dots,n+1)}_{(1,\dots,n)}) \quad = \quad n\Delta^{(1,\dots,\hat{l},\dots,n+1)}_{(1,\dots,n)}.$$

We are now able to compute the lambdas:

Proposition 9.2.4. In the basis $\underline{\omega}$ the matrix A_∞ is

$$A_\infty = \operatorname{diag}\begin{pmatrix} 0, & (n+1), & 2(n+1), & \dots & (n-1)(n+1), \\ n, & (n+1), & 2(n+1), & \dots & (n-1)(n+1), \\ 2n, & (n+1), & 2(n+1), & \dots & (n-1)(n+1), \\ \vdots & \vdots & \vdots & \dots & \vdots \\ n^2, & (n+1), & 2(n+1), & \dots & (n-1)(n+1) \end{pmatrix} \tag{9.5}$$

Proof. The claim follows from corollary 9.2.3 by the following calculation:

$$\begin{aligned} \lambda_i \cdot h_{n+1}\prod_{j=1}^{i-1} &= \theta_i(h_{n+1}\prod_{j=1}^{i-1} h_j \\ &= n\cdot h_{n+1}\prod_{j=1}^{i-1} h_j + h_{n+1}\sum_{l=1}^{i-1}\theta_i(h_l)\prod_{j\neq l} h_l \\ &= in\cdot h_{n+1}\prod_{j=1}^{i-1}. \end{aligned}$$

□

9.3 The spectrum of the star quiver

In this section we will prove a formula for the spectrum of the star quiver $*_n$.

We denote by $\mu = \dim(\operatorname{Rep}(*_n, \mathbf{d})$ and by $\underline{\omega}^{(1)}$ the basis from the previous section. We know from the corollary 9.1.8 and proposition 9.2.4 that in the basis $\underline{\omega}^{(1)}$ the connection is given by the matrix

$$\tau\partial_\tau\underline{\omega}^{(1)} = \underline{\omega}^{(1)}(A_0(t) + A_\infty)d\tau, \tag{9.6}$$

where the matrices are

$$A_0 = \begin{pmatrix} 0 & 0 & \dots & 0 & t \\ 1 & 0 & \dots & 0 & 0 \\ 0 & 1 & \dots & 0 & 0 \\ \dots & \dots & \dots & \dots & \\ 0 & 0 & \dots & 1 & 0 \end{pmatrix},$$

$$A_\infty = \operatorname{diag}\begin{pmatrix} 0, & (n+1), & 2(n+1), & \dots & (n-1)(n+1), \\ n, & (n+1), & 2(n+1), & \dots & (n-1)(n+1), \\ 2n, & (n+1), & 2(n+1), & \dots & (n-1)(n+1), \\ \vdots & \vdots & \vdots & \dots & \vdots \\ n^2, & (n+1), & 2(n+1), & \dots & (n-1)(n+1) \end{pmatrix}.$$

We know by proposition 2.2.5 that the spectrum at t is given by the diagonal entries of the matrix $A_\infty = \operatorname{diag}(\nu_1, \dots, \nu_\mu)$, when these entries fulfil the condition $\nu_i - \nu_{i-1} \leq 1$ for $i = 2, \dots, \mu$ for $t = 0$ and for $t \neq 0$ they moreover fulfil the condition $\nu_1 - \nu_\mu \leq 1$. Obviously these conditions are not fulfilled at the moment.
But by lemma 2.2.6 there exists a base change to a basis $\underline{\omega}^{(2)}$ (resp. $\underline{\omega}^{(3)}$), such that in the basis $\underline{\omega}^{(2)}$ (resp. $\underline{\omega}^{(3)}$) the entries of A_∞ fulfil the conditions for $t = 0$ (resp. $t \neq 0$) of proposition 2.2.6. The base changes are done by applying two algorithms, which can be formulate for the entries of the matrix A_∞ as follows:
Algorithm 1: Let $\underline{\nu}^{(1)}$ be the entries of the matrix A_∞ in the basis $\underline{\omega}^{(1)}$.
For $i = 2, \dots, \mu$ do :
If $\nu_i^{(1)} - \nu_{i-1}^{(1)} > 1$, then put $\tilde{\nu}_{i-1}^{(1)} = \nu_i^{(1)} - 1$, $\tilde{\nu}_i^{(1)} = \nu_{i-1}^{(1)} + 1$ and $\tilde{\nu}_j^{(1)} = \nu_j^{(1)}$ for $j \neq i, i-1$ and restart Algorithm 1 with input $\underline{\tilde{\nu}}^{(1)}$.

Remark 9.3.1. Of course the precise output of Algorithm 1 is a basis $\underline{\omega}^{(2)}$, such that the diagonal entries $\underline{\nu}^{(2)}$ of the corresponding matrix A_∞ fulfil the conditions for $t = 0$. But we will not need a precise formula for the base change. Hence it is much simpler to consider Algorithm 1 as an operation on diagonal entries of the matrix A_∞.
In other words Algorithm 1 does in each step the following :

$$\text{If } \nu_i - \nu_{i-1} > 1, \text{ then } \begin{pmatrix} \nu_{i-1} \\ \nu_i \end{pmatrix} \mapsto \begin{pmatrix} \nu_i - 1 \\ \nu_{i-1} + 1 \end{pmatrix}. \qquad \text{(ALG1)}$$

Algorithm 2: Let $\underline{\nu}^{(2)}$ be the entries of the matrix A_∞ in the basis $\underline{\omega}^{(2)}$.
If $\nu_1^{(2)} - \nu_\mu^{(2)} > 1$, then put $\tilde{\nu_1}^{(2)} = \nu_\mu^{(2)} + 1$, $\tilde{\nu}_\mu^{(2)} = \nu_1^{(2)} - 1$ and $\tilde{\nu}_j^{(2)} = \nu_j^{(2)}$ for $j \neq 1, \mu$. Then start Algorithm 1 with $\underline{\tilde{\nu}}^{(2)}$.

Remark 9.3.2. Again the precise output of Algorithm 2 is a basis $\underline{\omega}^{(3)}$, in which the diagonal entries $\underline{\nu}^{(3)}$ of the corresponding matrix A_∞ fulfil the conditions for $t \neq 0$, but again we will consider it by abuse of notation as an operation on the matrix A_∞ doing in its first step

$$\text{If } \nu_1 - \nu_\mu > 1, \text{ then } \begin{pmatrix} \nu_1 \\ \nu_\mu \end{pmatrix} \mapsto \begin{pmatrix} \nu_\mu + 1 \\ \nu_1 - 1 \end{pmatrix}. \qquad \text{(ALG2)}$$

Remark 9.3.3. If it is clear from the context, we are using a simplified notation and drop the upper index, i.e. $\nu_j = \nu_j^{(i)}$. Otherwise $\underline{\nu}^{(i)}$ denotes the diagonal entries of the matrix A_∞ in the basis $\underline{\omega}^{(i)}$ for $i = 1, 2, 3$. We also remark that it is enough to run Algorithm 2 just once, i.e. after shifting and interchanging the first and the last entry of A_∞ as above, the result of Algorithm 1 satisfies both, $\nu_i - \nu_{i-1} \leq 1$ for $i = 2, \ldots, \mu$ and $\nu_1 - \nu_\mu \leq 1$.

Before we study the general case, let us consider the example $*_5$. We start with the basis $\underline{\omega}^{(1)}$ from the previous section and write the diagonal entries of the matrix A_∞ in a 5×6 table, where the numbering of the cells is given by the following table

$$I = \begin{array}{|c|c|c|c|c|} \hline 1 & 2 & 3 & 4 & 5 \\ \hline 6 & 7 & 8 & 9 & \ldots \\ \hline \ldots & \ldots & \ldots & \ldots & \ldots \\ \hline \ldots & \ldots & \ldots & 24 & 25 \\ \hline 26 & 27 & 28 & 29 & 30 \\ \hline \end{array} . \tag{9.7}$$

Hence the table of the matrix A_∞ is

$$A_\infty = \begin{array}{|c|c|c|c|c|} \hline 0 & 6 & 12 & 18 & 24 \\ \hline 5 & 6 & 12 & 18 & 24 \\ \hline 10 & 6 & 12 & 18 & 24 \\ \hline 15 & 6 & 12 & 18 & 24 \\ \hline 20 & 6 & 12 & 18 & 24 \\ \hline 25 & 6 & 12 & 18 & 24 \\ \hline \end{array} . \tag{9.8}$$

According to the rule (ALG1) the table of the matrix A_∞ after the fist step of Algorithm 1 is

$$A_\infty = \begin{array}{|c|c|c|c|c|} \hline 5 & 1 & 12 & 18 & 24 \\ \hline 5 & 6 & 12 & 18 & 24 \\ \hline 10 & 6 & 12 & 18 & 24 \\ \hline 15 & 6 & 12 & 18 & 24 \\ \hline 20 & 6 & 12 & 18 & 24 \\ \hline 25 & 6 & 12 & 18 & 24 \\ \hline \end{array} \tag{9.9}$$

and after the second step it is

$$A_\infty = \begin{array}{|c|c|c|c|c|}\hline 5 & 11 & 2 & 18 & 24 \\ \hline 5 & 6 & 12 & 18 & 24 \\ \hline 10 & 6 & 12 & 18 & 24 \\ \hline 15 & 6 & 12 & 18 & 24 \\ \hline 20 & 6 & 12 & 18 & 24 \\ \hline 25 & 6 & 12 & 18 & 24 \\ \hline \end{array}. \tag{9.10}$$

The output of Algorithm 1 is the following table

$$A_\infty = \begin{array}{|c|c|c|c|c|}\hline 20 & 16 & 17 & 13 & 14 \\ \hline 15 & 11 & 12 & 13 & 14 \\ \hline 10 & 11 & 12 & 13 & 14 \\ \hline 15 & 16 & 17 & 18 & 19 \\ \hline 15 & 16 & 17 & 18 & 14 \\ \hline 15 & 16 & 12 & 13 & 9 \\ \hline \end{array}. \tag{9.11}$$

We remark that during Algorithm 1 each entry $\nu_i^{(1)}$ is moved r_i places to the left and s_i places to the right, according to the ordering given in (9.7), and $s_i - r_i$ is added.
We can obtain this result differently. Therefore we consider the table

$$\kappa_5 = \begin{array}{|c|c|c|c|c|}\hline 11 & 7 & 4 & 2 & 1 \\ \hline 12 & 13 & 8 & 5 & 3 \\ \hline 14 & 24 & 15 & 9 & 6 \\ \hline 16 & 27 & 23 & 17 & 10 \\ \hline 18 & 29 & 26 & 22 & 19 \\ \hline 20 & 30 & 28 & 25 & 21 \\ \hline \end{array} \tag{9.12}$$

as a permutation, i.e. the first cell goes to the 11 position, the second cell goes to the seventh position and so on. We use the convention that A_{κ_5} denotes the output of Algorithm 1 on a 5×6 tableau A and A^{κ_5} denotes the permutation of the entries of A according to κ_5. It is easy to see that

$$(A_\infty + \kappa_5 - I)^{\kappa_5} = \begin{array}{|c|c|c|c|c|}\hline 10 & 11 & 13 & 16 & 20 \\ \hline 11 & 12 & 12 & 14 & 17 \\ \hline 13 & 18 & 14 & 13 & 15 \\ \hline 15 & 16 & 17 & 16 & 14 \\ \hline 17 & 13 & 15 & 16 & 18 \\ \hline 19 & 9 & 12 & 14 & 15 \\ \hline \end{array}^{\kappa_5}$$

$$= \begin{array}{|c|c|c|c|c|}\hline 20 & 16 & 17 & 13 & 14 \\ \hline 15 & 11 & 12 & 13 & 14 \\ \hline 10 & 11 & 12 & 13 & 14 \\ \hline 15 & 16 & 17 & 18 & 19 \\ \hline 15 & 16 & 17 & 18 & 14 \\ \hline 15 & 16 & 12 & 13 & 9 \\ \hline \end{array}$$

$$= (A_\infty)_{\kappa_5}$$

The permutation κ_5 is constructed, like this

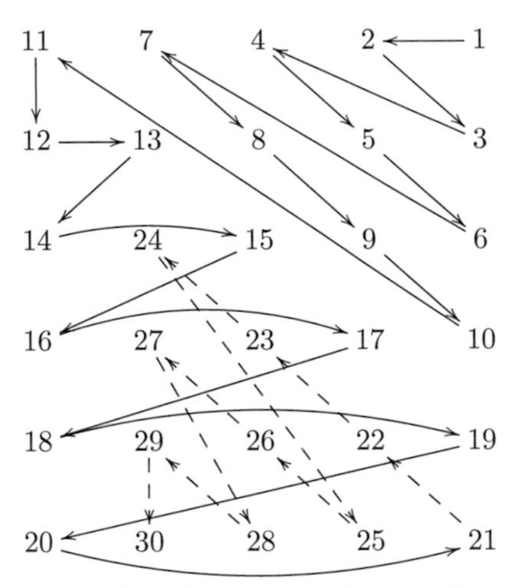

We observe that concerning the permutation κ_5 the entries of A_∞ can be roughly divided into the following different cases:

- entries above the diagonal, but not in the last column
- entries above the diagonal and in the last column
- entries on the diagonal

- entries below the diagonal, but not in the first column and not in the last row
- entries below the diagonal, but not in the first column and in the last row
- entries below the diagonal and in the first column

Before we deal with the general case, we recall that the spectral numbers $\underline{\nu}^{(i)}$, $i = 2, 3$, come in blocks and they are symmetric with respect to $n(n+1) - 1 = \dim(\mathrm{Rep}(*_n, \mathbf{d}) - 1$.

We arrange the spectral numbers at $t = 0$ $\nu_i^{(2)}$ with respect to the blocks and arrange the diagonal entries $\nu_i^{(1)}$ of the matrix A_∞ in the basis $\underline{\omega}^{(1)}$ in the same way. In our example $*_5$ we obtain the following picture:

20					0				
16	17				6	12			
13	14	15			18	24	5		
11	12	13	14		6	12	18	24	
10	11	12	13	14	10	6	12	18	24
15	16	17	18	19	15	6	12	18	24
15	16	17	18		20	6	12	18	
14	15	16			24	25	6		
12	13				12	18			
9					24				

where the $\nu_i^{(2)}$ are arranged on the left and the $\nu_i^{(1)}$ are arranged on the right.

We are now able to deal with the general case, i.e. we consider the tableau

$A_\infty =$

0	$n+1$	$2(n+1)$	$3(n+1)$	$\dots$	$\dots$	$(n-1)(n+1)$
n	$n+1$	$2(n+1)$	$3(n+1)$	$\dots$	$\dots$	$(n-1)(n+1)$
$2n$	$n+1$	$2(n+1)$	$3(n+1)$	$\dots$	$\dots$	$(n-1)(n+1)$
$\dots$	$\dots$	$\dots$	$\dots$	$\dots$	$\dots$	$\dots$
n^2	$n+1$	$2(n+1)$	$3(n+1)$	$\dots$	$\dots$	$(n-1)(n+1)$

and the permutation κ, given by generalizing the permutation κ_5. We denote by a_{ij} the entry of the $i-$row and the $j-$th column of the tableau A_∞. Now the action of κ_n translates to:

Lemma 9.3.4. If the entry a_{ij} is above the diagonal, then its moved by κ_n

$i(j-i-1)+\frac{(i+1)i}{2}$ positions to the left and

$\frac{(n-j+1)(n-j)}{2}$ positions to the right,

i.e. for $1 \leq i < j$ the total shift is of a_{ij} is

$$\mathrm{shift}(i,j) = \frac{1}{2}n^2 + \frac{1}{2}i^2 - nj - ij + \frac{1}{2}j^2 + \frac{1}{2}n + \frac{1}{2}i - \frac{1}{2}j.$$

Lemma 9.3.5. If the entry a_{ij} is below the diagonal and not in the first column, then its moved by κ_n

$\frac{j(j-1)}{2} - 1$ positions to the left and

$(n-i+2)(i-j-1) - 1 + \frac{(n+3-i)(n+2-i)}{2} + (n+1-i)$ positions to the right,

i.e. for $2 \leq j < i$ the total shift of a_{ij} by κ_n is

$$\mathrm{shift}(i,j) = \frac{1}{2}n^2 - \frac{1}{2}i^2 - nj + ij - \frac{1}{2}j^2 + \frac{5}{2}n - \frac{1}{2}i - \frac{3}{2}j + 2.$$

Lemma 9.3.6. For the entries a_{i1} of the first column with $i \geq 2$ the total shift by κ_n is

$$\mathrm{shift}(i,1) = \frac{(n-i+1)(n-i)}{2} - \frac{(i-2)(i-3)}{2}.$$

Lemma 9.3.7. The total shift of an entry a_{ii} of the diagonal is given by

$$\mathrm{shift}(i,i) = \frac{(n-i+1)(n-i)}{2} - \frac{(i-1)(i-2)}{2}.$$

For $j > 1$ we have $a_{ij} = (j-1)(n+1)$ and in the first column we have $a_{i1} = (i-1)n$.

Proposition 9.3.8. After the base change corresponding to the permutation κ_n the entries of the matrix A_∞ fulfil the conditions

$$\nu_i^{(2)} - \nu_{i-1}^{(2)} \leq 1.$$

Proof. The value of the (i,j)–th entry in the tableau A_∞ is

$$(i,j) = \begin{cases} (j-1)(n+1), & \text{for } j \neq 1 \\ (i-1)n, & \text{for } j = 1 \end{cases}.$$

By the shape of the permutation κ_n we have to consider the following cases:

1. For an entry above the diagonal and in the last column the successor of (i, n) is $(1, n-i)$.

$$[(1, n-i) + \text{shift}(1, n-i)] - [(i, n) + \text{shift}(i, n)] = 1 - n$$

2. For an entry above the diagonal and not in the last column the successor of (i, j) is $(i+1, j+1)$.

$$[(i+1, j+1) + \text{shift}(i+1, j+1)] - [(i, j) + \text{shift}(i, j)] = 1.$$

3. An entry of the diagonal (i, i) is successed by element $(i+1, 1)$ of the first column.

$$[(i+1, 1) + \text{shift}(i+1, 1)] - [(i, i) + \text{shift}(i, i)] = 1.$$

4. An entry $(i, 1)$, $i \neq 1, n+1$, of the first column is successed by the entry (i, i).

$$[(i, i) + \text{shift}(i, i)] - [(i, 1) + \text{shift}(i, 1)] = 1.$$

5. (Special case) The entry $(n+1, 1)$ is successed by the entry $(n+1, n)$.

$$[(n+1, n) + \text{shift}(n+1, n)] - [(n+1, 1) + \text{shift}(n+1, 1)] = 1 - n.$$

6. For an entry below the diagonal and in the second column the successor of $(n-i+2, 2)$ is $(n+1, i)$.

$$[(n-i+2, 2) + \text{shift}(n-i+2, 2)] - [(n+1, i)\text{shift}(n+1, i)] = 1 - n.$$

7. For an entry below the diagonal, which is not in the second column, the successor of (i, j) is $(i+1, j+1)$.

$$[(i+1, j+1) + \text{shift}(i+1, j+1)] - [(i, j) + \text{shift}(i, j)] = 1.$$

□

Hence the computation of the spectral numbers at $t = 0$ follows by induction from above starting with the first number which is equal to $(1, n) + \text{shift}\,(1, n) = n(n-1)$

Corollary 9.3.9. The spectrum of $*_n$ at $t = 0$ is

$n(n-1)$					
$(n-1)^2$	$(n-1)^2+1$				
$(n-2)(n-1)+1$	$(n-2)(n-1)+2$	$(n-2)(n-1)+3$			
...					
$n(n-1)/2$	$n(n-1)/2+1$	$n(n-1)/2+2$	$\cdots$	$(n+2)(n-1)/2-1$	$(n+2)(n-1)/2$
$(n+2)(n-1)/2+1$	$(n+2)(n-1)/2+2$	$(n+2)(n-1)/2+3$	$\cdots$	$(n+4)(n-1)/2$	$(n+4)(n-1)/2+1$
...					
$4n-6$	$4n-5$	$4n-4$			
$3n-3$	$3n-2$				
$2n-1$					

The procedure from above works for $t = 0$, but in the case $t \neq 0$ we can not conclude that those numbers are the spectral numbers, because in general the last one is much smaller than the first one. The only cases, where the spectral numbers coincide for any t are $n = 2, 3$. However it is possible to obtain the spectral numbers for $t \neq 0$ from those at $t = 0$. We illustrate this with our example $*_5$.
We start from table (9.11), which we write as follows:

20				
16	17			
13	14	15		
11	12	13	14	
10	11	12	13	14
15	16	17	18	19
15	16	17	18	
14	15	16		
12	13			
9				

Then we apply Algorithm 2, i.e. we carry out the following changes:

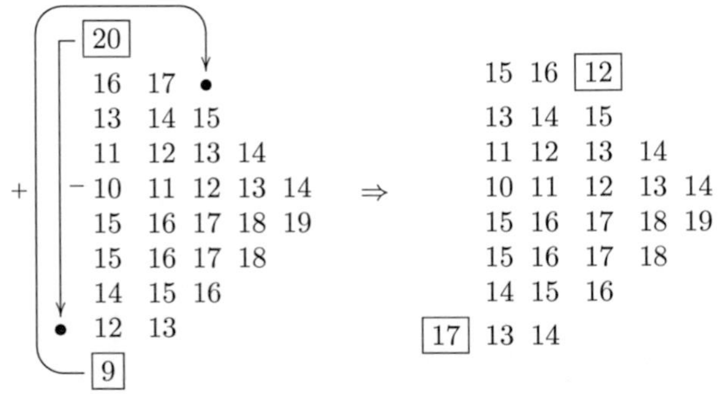

$$
\begin{array}{l}
\boxed{20} \\
16\ \ 17\ \ \bullet \\
13\ \ 14\ \ 15 \\
11\ \ 12\ \ 13\ \ 14 \\
10\ \ 11\ \ 12\ \ 13\ \ 14 \\
15\ \ 16\ \ 17\ \ 18\ \ 19 \\
15\ \ 16\ \ 17\ \ 18 \\
14\ \ 15\ \ 16 \\
\bullet\ \ 12\ \ 13 \\
\boxed{9}
\end{array}
\quad \Rightarrow \quad
\begin{array}{l}
15\ \ 16\ \ \boxed{12} \\
13\ \ 14\ \ 15 \\
11\ \ 12\ \ 13\ \ 14 \\
10\ \ 11\ \ 12\ \ 13\ \ 14 \\
15\ \ 16\ \ 17\ \ 18\ \ 19 \\
15\ \ 16\ \ 17\ \ 18 \\
14\ \ 15\ \ 16 \\
\boxed{17}\ \ 13\ \ 14
\end{array}
\tag{9.13}
$$

(In the left diagram, arrows marked "+" and "−" indicate the moves.)

In the general case the tableau is

$(n-1)^2-1$	$(n-1)^2$	
$(n-2)(n-1)$	$(n-2)(n-1)+1$	$(n-2)(n-1)+2$

$\cdots\cdots\cdots\cdots\cdots\cdots\cdots\cdots\cdots\cdots\cdots\cdots\cdots\cdots$

$n(n-1)/2+2$	$n(n-1)/2+3$	$n(n-1)/2+4$	$\cdots$	$n(n+1)/2$	
$n(n-1)/2+1$	$n(n-1)/2+2$	$n(n-1)/2+3$	$\cdots$	$n(n+1)/2-1$	
$n(n-1)/2$	$n(n-1)/2+1$	$n(n-1)/2+2$	$\cdots$	$(n+2)(n-1)/2-1$	$(n+2)(n-1)/2$
$(n+2)(n-1)/2+1$	$(n+2)(n-1)/2+2$	$(n+2)(n-1)/2+3$	$\cdots$	$(n+4)(n-1)/2$	$(n+4)(n-1)/2+1$
$n(n+1)/2$	$n(n+1)/2+1$	$n(n+1)/2+2$	$\cdots$	$n(n+3)/2-2$	
$n(n+1)/2-1$	$n(n+1)/2$	$n(n+1)/2+1$	$\cdots$	$n(n+3)/2-3$	

$\cdots\cdots\cdots\cdots\cdots\cdots\cdots\cdots\cdots\cdots\cdots\cdots\cdots\cdots$

$4n-6$	$4n-5$	$4n-4$
$3n-3$	$3n-2$	
$2n-1$		

and we need to permute the first and the last element as indicated in (9.13). After interchanging, we move the last element $1+2+\ldots+(n-3) = (n-2)(n-3)/2$ positions to the right. Then its value changes to

$$2n-1 \mapsto (2n-1)+1+(n-2)(n-3)/2 = n(n+1)/2+2.$$

The next element is then $(1,3)+\text{shift}(1,3) = n(n-1)/2+3$, so that the jump is still equal to 1. Also the previous element is $(n-3,n)+\text{shift}(n-3,n)-1 = n(n+1)/2+1$, so that the jump is $1-n$. The first element $n(n-1)$ we move $1+2+\ldots(n-3) = (n-2)(n-3)/2$ positions to the left. Then its value changes to

$$n(n-1) \mapsto n(n-1)-1-(n-2)(n-3)/2 = n(n+3)/2-3.$$

The previous element is then $(4,2)+\text{shift}(4,2) = n(n+3)/2-4$ and again the jump is equal to 1. We conclude:

Proposition 9.3.10. The spectral numbers of the star quiver $*_n$ at $t \neq 0$ are:

$(n-1)^2-1$	$(n-1)^2$	
$(n-2)(n-1)$	$(n-2)(n-1)+1$	$(n-2)(n-1)+2$

……

$n(n-1)/2+2$	$n(n-1)/2+3$	$n(n-1)/2+4$	$\cdots$	$n(n+1)/2$	
$n(n-1)/2+1$	$n(n-1)/2+2$	$n(n-1)/2+3$	$\cdots$	$n(n+1)/2-1$	
$n(n-1)/2$	$n(n-1)/2+1$	$n(n-1)/2+2$	$\cdots$	$(n+2)(n-1)/2-1$	$(n+2)(n-1)/2$
$(n+2)(n-1)/2+1$	$(n+2)(n-1)/2+2$	$(n+2)(n-1)/2+3$	$\cdots$	$(n+4)(n-1)/2$	$(n+4)(n-1)/2+1$
$n(n+1)/2$	$n(n+1)/2+1$	$n(n+1)/2+2$	$\cdots$	$n(n+3)/2-2$	
$n(n+1)/2-1$	$n(n+1)/2$	$n(n+1)/2+1$	$\cdots$	$n(n+3)/2-3$	

……

$4n-6$	$4n-5$	$4n-4$
$3n-3$	$3n-2$	
$2n-1$		

Applying to this result corollary 6.2.10, we obtain the roots of the Bernstein-Sato polynomial of $h = \prod h_i$:

Corollary 9.3.11. The Bernstein-Sato polynomial of the canonical equation Δ for the star quiver $*_n$ is

$$b_\Delta(s) = \prod_{k=0}^{n-2}(s+\frac{2n-k}{n+1})^{k+1}(s+1)^{2n}\prod_{l=0}^{n-2}(s+\frac{n-l}{n+1})^{n-l-1}.$$

9.4 The exponent condition and the invariant subspace conjecture for the star quiver

Proposition 9.4.1. The star quiver $*_n$ fulfils the exponent condition 8.1.4 and the invariant subspace conjecture 8.1.1.

Proof. We begin with the exponent condition 8.1.4. The Bernstein-Sato polynomial for the star quiver $*_n$ is by corollary 9.3.11

$$b_\Delta(s) = \prod_{k=0}^{n-2}(s+\frac{2n-k}{n+1})^{k+1}(s+1)^{2n}\prod_{l=0}^{n-2}(s+\frac{n-l}{n+1})^{n-l-1}.$$

It is obvious that we only have to deal with the roots

$$(-\frac{2n-k}{n+1})^{k+1} \quad , \quad \text{for } k=0,\ldots,n-2$$
$$(-\frac{n-l}{n+1})^{n-l-1} \quad , \quad \text{for } l=0,\ldots,n-2,$$

where the exponent denotes the multiplicity of the root.
After shifting the roots $-\frac{n-l}{n+1}$, $l = 0, \dots, n-2$, by $+\frac{n+1}{n+1}$, we obtain

$$0 < \frac{n+1}{n+1} - \frac{n-l}{n+1} = \frac{l+1}{n+1} < 1$$

for $l = 0, \dots, n-2$. After shifting the roots $-\frac{2n-k}{n+1}$, $k = 0, \dots, n-2$, by $+\frac{2(n+1)}{n+1}$, we have

$$0 < -\frac{2n-k}{n+1} + \frac{2(n+1)}{n+1} = \frac{k+2}{n+1} < 1$$

for $k = 0, \dots, n-2$.
Obviously every fraction $\frac{i}{n+1}$, $i = 1, \dots, n$, occurs in the tuple

$$R = \left((\frac{k+2}{n+1})^{k+1}, (\frac{l+1}{n+1})^{n-l-1} \right)_{k,l=0,\dots,n-2},$$

where the exponents denote the multiplicity. Moreover each component of the tuple R is a fraction of the form $\frac{i}{n+1}$ for some $i = 1, \dots, n$.
We claim that every fraction $\frac{i}{n+1}$, $i = 1, \dots, n$ occurs with multiplicity $n-1$ in R. Obviously we have

$$\frac{k+2}{n+1} = \frac{l+1}{n+1} \Leftrightarrow k+1 = l$$

and in this case the exponent of the shifted root $\frac{i}{n+1}$, $i = 2, \dots, n-1$, is equal to $k+1+n-l-1 = n-1$. So it remains to look at the exponents of the shifted roots $\frac{k+1}{n+1}$ for $k = n-2$ and $\frac{l+1}{n+1}$ for $l = 0$. As in the remaining cases $k = n-2$ and $l = 0$ the exponents are $n-1$ too, we conclude that the star quiver $*_n$ fulfils the exponent condition 8.1.4.

For the invariant subspace conjecture 8.1.1 we first note that the quiver group of $*_n$ with dimension vector $\mathbf{d} = (n, 1, \dots, 1)$ is

$$\mathrm{GL}(*_n, \mathbf{d}) = \mathrm{GL}_n(\mathbb{C}) \times \prod_{i=1}^{n+1} \mathrm{GL}_1(\mathbb{C})$$

and the maximal torus given by the diagonal matrices is

$$T = \prod_{i=1}^{n} \mathrm{GL}_1(\mathbb{C}) \times \prod_{i=1}^{n+1} (\mathbb{C}).$$

We denote by $\tilde{T}$ the image of T under the canonical projection $\pi\colon \mathrm{GL}(*_n\mathbf{d}) \rightarrow \mathrm{PGL}(*, \mathbf{d})$. It is clear that $\tilde{T}$ is a maximal torus of $\mathrm{PGL}(*_n, \mathbf{d})$ of dimension

$$\dim(\tilde{T}) = 2n.$$

So we conclude from corollary 9.3.11 that the condition $(C3)$ is satisfied. Let e_i denote the $i-$th standard basis vector of $\mathbb{C}^n$. We consider the generic representation

$$M = (e_1, \ldots, e_n, \sum_{i=1}^{n} e_i).$$

One easily calculates that the closure of the orbit $\tilde{T}.M$ consists of matrices of the form

$$\begin{pmatrix} * & 0 & \ldots & 0 & * \\ 0 & * & \ldots & 0 & * \\ \ldots & \ldots & \ldots & \ldots & \ldots \\ 0 & 0 & \ldots & * & * \end{pmatrix},$$

where each entry marked by $*$ is a complex number. We see that the condition $(C1) : \dim(\tilde{T}) = \dim(W_T)$ is obviously satisfied. But this also follows in general: As M is a generic representation, i.e. its stabilizer in T and in $\mathrm{GL}(*_n, \mathbf{d})$ is isomorphic to $\mathbb{C}^*$, we conclude from the equality $\dim(T.M) = \dim(T) - dim(T_M)$ that $\dim(W_T) = \dim(\tilde{T})$.
Before we prove the condition $(C2)$, we recall that the discriminant of the pair $(\mathrm{GL}(*_n, \mathbf{d}), \mathrm{Rep}(*, \mathbf{d}))$ is defined by $h = \prod_{i=1}^{n+1} h_i$, where h_i was the determinant of the maximal minor of the $n \times (n+1)$ matrix $\Delta = (x_{ij})_{ij}$ obtained by deleting the $i-$th column.
It follows that the restriction of h to W_T is a monomial. Indeed up to a sign the restrictions of the h_i to W_T are

$$\begin{aligned} (h_{n+1})_{|W_T} &= \prod_{i=1}^{n} x_{ii} \\ (h_j)_{|W_T} &= x_{j(n+1)} \prod_{i=1, i \neq j}^{n} x_{ii}, \text{ for } j = 1, \ldots, n. \end{aligned}$$

We conclude that the restriction h_{W_T} is a monomial and the condition $(C2)$ is satisfied.

□

Chapter 10

The serie $\mathrm{Ext}_k\mathrm{Star}_n$

In this chapter we describe a simple construction with quivers and dimension vectors, which we can use to obtain series of examples of reductive regular prehomogeneous vector spaces. Moreover it will become clear that many of the previous examples are related by this construction. Moreover we will apply this construction to the star quiver and calculate the spectrum of the FL-Brieskorn lattice and the Bernstein-Sato polynomials of defining equations of the discriminants.

10.1 Constructing new Series

We will now describe a simple construction with quivers and dimension vectors. This construction will have the property that a quiver with a prehomogeneous dimension vector gives rise to a new quiver with new prehomogeneous dimension vector.

Let $Q = (Q_0, Q_1)$ be a quiver and $x \in Q_0$ be a node. We denote by $\tilde{I}_x := \{\alpha \in Q_1 \mid t\alpha = x \text{ or } h\alpha = x\} \subset Q_1$ the subset of arrows, whose tail or head is the node x. For a subset $I_x \subset \tilde{I}_x$ we construct a new quiver $Q^{I_x} = (Q_0^{I_x}, Q_1^{I_x})$ as follows:

First we replace the node x by a pair of nodes x', x''. Next we connect x' and x'' by an arrow[1] $x' \to x''$ and at last we attach the arrows of I_x

[1]Changing the direction of this arrow may have an impact on the equations of the irreducible components of the discriminant later, but at the moment we restrict ourself to this case for simplicity.

to x' and of $\tilde{I}_x \backslash I_x$ to x''. For example let us consider the quiver

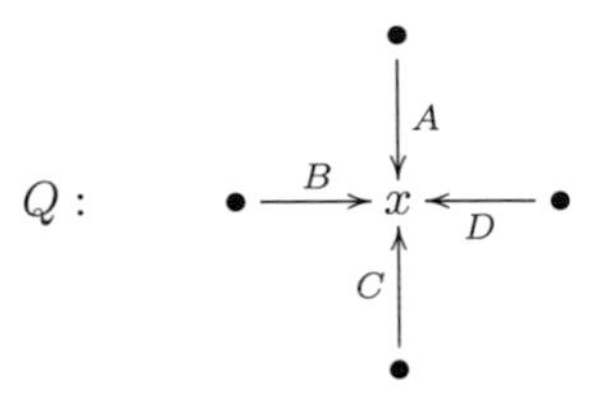

Then two possible outcomes of this process for Q are

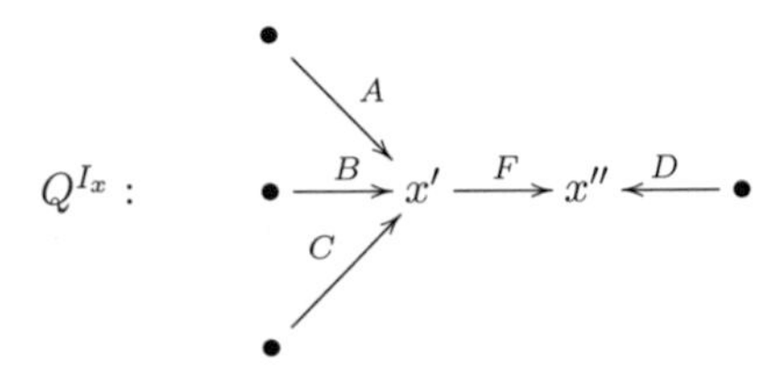

for $I_x = \{A, B, C\}$, and

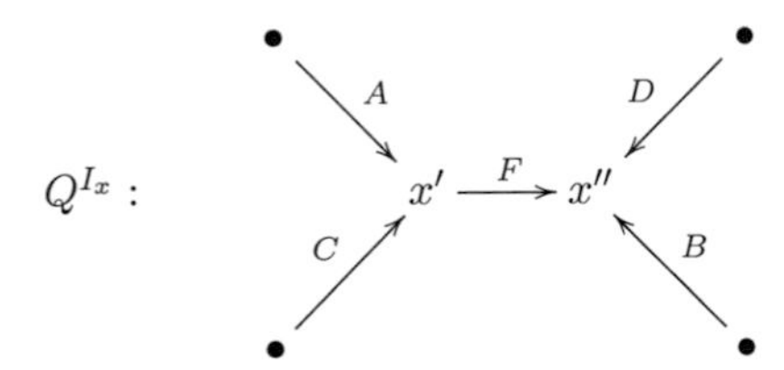

for $I_x = \{A, C\}$. We also allow the cases $I_x = \emptyset$ and $I_x = \tilde{I}_x$ and we will usually identify the set Q_1 with a subset of $Q_1^{I_x}$.
Let $\mathbf{d} \in \mathbb{N}^{Q_0}$ be a dimension vector. We define a dimension vector $\mathbf{d}_x \in \mathbb{N}^{Q_0^{I_x}}$ by setting $\mathbf{d}_x(i) = \mathbf{d}(i)$ for all $i \in Q_0^{I_x} \backslash \{x' x''\}$ and $\mathbf{d}_x(x') = \mathbf{d}_x(x'') = \mathbf{d}(x)$.
Moreover given a representation $M = (V(\alpha))_{\alpha \in Q_1} \in \operatorname{Rep}(Q, \mathbf{d})$ and an endomorphism $\phi \in \operatorname{End}(k^{\mathbf{d}_x})$, we define the representation $M_\phi^{I_x} = (V_\alpha^{I_x}) \in \operatorname{Rep}(Q^{I_x}, \mathbf{d}_x)$ by $V_\phi^{I_x}(\alpha) = V(\alpha)$ for all $\alpha \in Q_1$ and $V(x' \to x'') := \phi$.
In sequel we will mainly consider the special case $\phi = \mathrm{id}$. Hence we set $M^{I_x} := M_{\mathrm{id}}^{I_x}$.

Remark 10.1.1. We remark that this construction was used in [BM06] example 7.7 to produce examples of linear free divisors. But in general

this construction does not give a linear free divisor, if it is applied to a pair $(Q, \mathbf{d})$, which defines a linear free divisor.
Let us understand why this is so, we recall first: If (G, V) is a prehomogeneous vector space, such that G is a connected subgroup of $\mathrm{GL}(V)$ with $\dim(G) = \dim(V)$ and $\Delta := \det(A_1.x| \ \ldots \ |A_n.x)$ is a reduced equation for the complement D of the open orbit, where $A_1, \ldots, A_n$ is a basis of the Lie algebra $\mathfrak{g}$ of G, then by remark 1.4.2 D is linear free divisor. And conversely we can associate to any linear free divisor $D \subset V$ such a pair (G, V).
Now let $(Q, \mathbf{d})$ be a quiver with real Schur root $\mathbf{d}$, such that $q_Q(\mathbf{d}) = 1$, then by corollary 7.2.16 and theorem 7.2.20 the pair $(\mathrm{PGL}(Q, \mathbf{d}), \mathrm{Rep}(Q, \mathbf{d}))$ is a prehomogeneous vector space and the complement D of the open orbit is a divisor. Hence such a pair $(Q, \mathbf{d})$ defines a linear free divisor if and only if Δ is reduced.
Now let us come back to our construction. The problem in general is that the equation Δ of the new pair $(\mathrm{PGL}(Q^{I_x}, \mathbf{d}_x), \mathrm{Rep}(Q^{I_x}, \mathbf{d}_x))$ is not reduced. In order to see this (cf. [BM06] example 7.7) we consider star quiver $*_4$ with dimension vector $\mathbf{d} = (4, 1, \ldots, 1)$

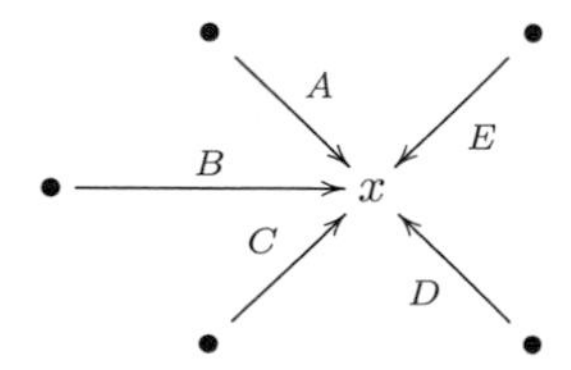

then two possible outcomes of the construction are $(Q1)$

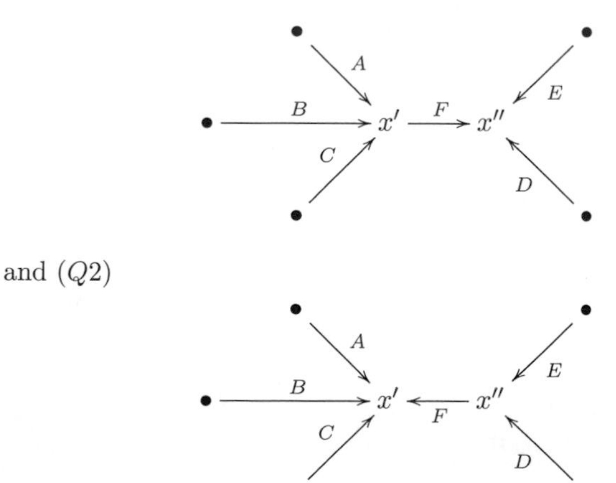

and $(Q2)$

Now the discriminant of $(Q1)$ is given by the vanishing of

$$\begin{array}{ccc} \det(FA|FB|FC|D), & \det(FA|FB|FC|E), & \det(FA|FB|D|E), \\ \det(FA|FC|E|D), & \det(FB|FC|E|D), & \det(F) \end{array}$$

in $\mathrm{Rep}(Q_1, \mathbf{d}_x)$ and the discriminant of $(Q2)$ is given by the vanishing of

$$\begin{array}{ccc} \det(A|B|C|FD), & \det(A|B|C|FE), & \det(A|B|FD|FE), \\ \det(A|C|FD|FE), & \det(B|C|FD|FE), & \det(F) \end{array}$$

in $\mathrm{Rep}(Q_2, \mathbf{d}_x)$. Now the degrees of the reduced equations for the discriminants are 36 for $(D1)$ and 32 for $(D2)$. As $\dim(\mathrm{Rep}(Q_i, \mathbf{d}_x) = 36 = \deg(\Delta)$ only the first one is a linear free divisor.

Fortunately we can also deal with non reduced equations, hence by the following theorem we are able (in principle) to produce lots of new examples beside the serie we introduce in the next section and most of the examples discussed in this thesis, can be obtained by applying this construction.

Proposition 10.1.2. If a representation $M \in \mathrm{Rep}(Q, \mathbf{d})$ is a brick, then the representation $M^{I_x} \in \mathrm{Rep}(Q^{I_x}, \mathbf{d}_x)$ is a brick.
In particular a real Schur root $\mathbf{d}$ is send to a real Schur root $\mathbf{d}_x$.

Proof. We index the nodes of Q by natural numbers, i.e. $Q_0 = (1, , \ldots, n)$, and recall that by definition 7.2.10 a representation $M \in \mathrm{Rep}(Q, \mathbf{d})$ is a brick if and only if $\mathrm{End}_Q(M) \cong \mathbb{C}$. Moreover this is equivalent to $\mathrm{Aut}_Q(M) \cong \mathbb{C}^*$ by remark 7.2.9.
Let $g = (g_1, \ldots, g_x, \ldots, g_n)$ be an element of $\mathrm{GL}(Q, \mathbf{d})$.Then g lies by definition in $\mathrm{Aut}_Q(V)$ if and only if $g.M = M$ holds.
Hence a representation $M = (V(\alpha))_{\alpha \in Q_1}$ is a brick, if and only if the system of equations

$$g_{h\alpha} V(\alpha) g_{t\alpha}^{-1} = V(\alpha), \ \alpha \in Q_1,$$

has precisely the solutions $\{g_\lambda = (\lambda \mathrm{id}_{\mathbf{d}_1}, \ldots, \lambda \mathrm{id}_{\mathbf{d}_n}) \mid \lambda \in \mathbb{C}^*\}$.
Let $g^x = (g_1^x, \ldots, g_{x'}^x, g_{x''}^x, \ldots, g_n^x) \in \mathrm{GL}(Q^{I_x}, \mathbf{d}_x)$ and let $M = (V(\alpha))_{\alpha \in Q_1} \in \mathrm{Rep}(Q, \mathbf{d})$ be a brick. We claim that $M^{I_x} = (V^{I_x}(\alpha))_{\alpha \in Q_1^{I_x}}$ is a brick. Therefore we denote the arrow $x' \to x''$ by ω and consider the system of equations

$$\begin{array}{rcl} g_{h\alpha}^x V^{I_x}(\alpha)(g_{t\alpha}^x)^{-1} & = & V^{I_x}(\alpha), \ \alpha \in Q_1, \\ g_{h\omega}^x V^{I_x}(\omega)(g_{t\omega}^x)^{-1} & = & V^{I_x}(\omega). \end{array}$$

As we have by construction $V^{I_x}(\omega) = \mathrm{id}$, we have

$$g^x_{x''} = g^x_{h\omega} = g^x_{t\omega} = g^x_{x'}.$$

Hence the system $g^x.M^{I_x} = M^{I_x}$ is equivalent to $g.M = M$. Indeed the system $g^x.M^{I_x} = M^{I_x}$ is

$$\begin{aligned}
g^x_{h\alpha}V(\alpha)(g^x_{t\alpha})^{-1} &= V(\alpha), \text{ for } \alpha \in Q_1 \backslash \tilde{I} \\
g^x_{h\alpha}V(\alpha)(g^x_{t\alpha})^{-1} &= V(\alpha), \text{ for } \alpha \in \tilde{I}, t\alpha = x', \alpha \neq \omega \\
g^x_{h\alpha}V(\alpha)(g^x_{t\alpha})^{-1} &= V(\alpha), \text{ for } \alpha \in \tilde{I}, h\alpha = x'', \alpha \neq \omega \\
g^x_{h\omega}V^{I_x}(\omega)(g^x_{t\omega})^{-1} &= V^{I_x}(\omega) = \mathrm{id}
\end{aligned}$$

because of $V^{I_x}(\alpha) = V(\alpha)$ for all $\alpha \in Q_1$.
We conclude from the assumption M is a brick that $g^x = (\lambda \mathrm{id}_{\mathbf{d}_1}, \ldots, \lambda \mathrm{id}_{\mathbf{d}_n})$ for $\lambda \in \mathbb{C}^*$ and hence M^{I_x} is a brick. □

We recall that the equality $q_Q(\mathbf{d}) = 1$ is equivalent to the equality $\dim(\mathrm{PGL}(Q, \mathbf{d})) = \mathrm{Rep}(Q, \mathbf{d}))$. By the following lemma the Tits form q_Q is invariant under this construction.

Lemma 10.1.3. We have $q_Q(\mathbf{d}) = q_{Q^{I_x}}(\mathbf{d}_x)$.

Proof. We recall that the Tits form of a dimension vector $\mathbf{d}$ of a quiver Q is defined as

$$q_Q(\mathbf{d}) = \sum_{i \in Q_0} \mathbf{d}(i)^2 - \sum_{\alpha \in Q_1} \mathbf{d}(t\alpha) \cdot \mathbf{d}(h\alpha).$$

We denote by ω the new arrow $x' \to x''$, i.e. $t\omega = x'$ and $h\omega = x''$, and conclude from the equalities $\mathbf{d}(i) = \mathbf{d}_x(i)$ for $i \in Q_0 \backslash \{x\}$ and $\mathbf{d}_x(x') = \mathbf{d}_x(x'') = \mathbf{d}(x)$:

$$\begin{aligned}
q_{Q^{I_x}}(\mathbf{d}_x) &= \sum_{i \in Q_0 \backslash \{x\}} \mathbf{d}_x(i)^2 + \mathbf{d}_x(x')^2 + \mathbf{d}_x(x'')^2 \\
&- \sum_{\alpha \in Q_1} \mathbf{d}_x(t\alpha)\mathbf{d}_x(h\alpha) + \mathbf{d}_x(t\omega)\mathbf{d}_x(h\omega) \\
&= (\sum_{i \in Q_0} (\mathbf{d}(i)^2 - \sum_{\alpha \in Q_1} \mathbf{d}(t\alpha)\mathbf{d}(h\alpha)) + \mathbf{d}(x)^2 - \mathbf{d}_x(x')\mathbf{d}_x(x'') \\
&= q_Q(\mathbf{d}).
\end{aligned}$$

□

We summarise the discussion:

Corollary 10.1.4. Let $(Q, \mathbf{d})$ define regular reductive prehomogeneous vector space. Then the pair $(Q^{I_x}, \mathbf{d}_x)$ defines an regular prehomogeneous vector space.

10.2 The fundamental semi-invariants of $\mathrm{Ext}_k\mathrm{Star}_n$

We recall that the symbol $\mathbb{C}^1 \overset{A_I}{\rightrightarrows} \mathbb{C}^n$ denoted the representation consisting of $\#I$ linear maps $A_i\colon \mathbb{C}^1 \to \mathbb{C}^n$, which correspond to $\#I$ arrows from $\#I$ outer nodes to one inner node.

If we apply the construction from the previous section to the inner node of the star quiver $*_n$ $k-$times, we obtain:

Definition 10.2.1. Let I and J be index sets of arrows with $\#I + \#J = n+1$. We denote by $\mathrm{Ext}_k\mathrm{Star}_n(I)$ the following pair of a quiver with dimension vector

$$\mathbb{C}^1 \overset{A_I}{\rightrightarrows} \mathbb{C}^n \overset{B_1}{\rightrightarrows} \dots \overset{B_k}{\rightrightarrows} \mathbb{C}^n \overset{A_J}{\leftleftarrows} \mathbb{C}^1.$$

Of course to be precise, the above picture shows a representation, but we can easily read of the picture the underlying quiver the and dimension vector. We put $B = \prod_{l=1}^k B_l$ and assume without loss of generality that $I = \{1, \dots, l\}$ and $J = \{l+1, \dots, n+1\}$.

Proposition 10.2.2. Let $l = \#I$. If $l \neq n+1, n$, then fundamental semi-invariants of $\mathrm{Ext}_k\mathrm{Star}_n(I)$ are

$$\begin{aligned} \tilde{h}_s &= \det(B_s),\ s = 1, \dots, k \\ h_i &= \det(BA_1|\dots|\widehat{BA_i}|\dots|BA_l|A_{l+1}|\dots|A_{n+1}),\ i \in I \\ h_j &= \det(BA_1|\dots|BA_l|A_{l+1}|\dots|\widehat{A_j}|\dots|A_{n+1}),\ j \in J. \end{aligned}$$

If $l = n$, then the fundamental semi-invariants are

$$\begin{aligned} \tilde{h}_s &= \det(B_s),\ s = 1, \dots, k \\ h_i &= \det(BA_1|\dots|\widehat{BA_i}|\dots|BA_n|A_{n+1}),\ i \in I \\ h_{n+1} &= \det(A_1|\dots|A_n). \end{aligned}$$

If $l = n+1$, then the fundamental semi-invariants are

$$\begin{aligned} \tilde{h}_s &= \det(B_s),\ s = 1, \dots, k \\ h_i &= \det(A_1|\dots|\widehat{A_i}|\dots|A_{n+1}),\ i \in I. \end{aligned}$$

Here $(A_1|\dots|\widehat{A_i}|\dots|A_{n+1})$ denotes the matrix, where we have removed the column A_i.

Proof. We first consider the case $l \neq n+1, n$, because the cases $l = n+1, n$ follow in a certain sense from this one.
It follows from a direct calculation that the functions h_s, h_j and h_i are semi-invariants. Obviously they are irreducible, if $l \neq n+1, n$. Hence they define the irreducible components of the discriminant and are fundamental semi-invariants. Last we conclude from Kac's result, see 7.3.8, that these are all fundamental semi-invariants.

If $\#J = 1$, then the semi-invariant

$$h_{n+1} = \det(BA_1|\dots|BA_n|\widehat{A_{n+1}}),$$

is reducible. Indeed we have

$$h_{n+1} = \det(BA_1|\dots|BA_n) = \det(B)\det(A_1|\dots|A_n),$$

but obviously $h_{n+1} = \det(A_1|\dots|A_n)$ is an irreducible semi-invariant, which finishes the proof in this case.
In the case $J = \emptyset$ the claim follows by the same argument. Indeed the semi-invariants

$$h_i = \det(BA_1|\dots|\widehat{BA_i}|\dots|BA_{n+1}), \text{ for } i \in I,$$

are reducible, but

$$h_i = \det(A_1|\dots|\widehat{A_i}|\dots|A_{n+1}), \text{ for } i \in I,$$

are irreducible semi-invariants. This finishes the proof. □

Proposition 10.2.3. Let $k > 0$, then the canonical equation of the discriminant of $\mathrm{Ext}_k\mathrm{Star}_n(I)$ is reduced if and only if $\#I = l = n, n-1$.

Proof. Let $\#I = l$ and assume $\#J \neq 1$, i.e. $\#I \neq n$. First we remark that for $\mathrm{Ext}_k\mathrm{Star}_n(I)$ the dimension of the representation space is

$$\mathrm{Rep}(\mathrm{Ext}_k\mathrm{Star}_n(I) = (n+1)\cdot n + k \cdot n^2.$$

We put $B := \prod_{j=k}^{1} B_j$ and distinguish the equations of the irreducible components of the discriminant as follows:

$$\begin{array}{rcl}
(Type\ 1) & & i = 1, \dots, l \\
h_i & = & \det(B\cdot A_1|\dots|\widehat{B\cdot A_i}|\dots|B\cdot A_l|A_{l+1}|\dots|A_{n+1}) \\
(Type\ 2) & & i = l+1, \dots, n+1 \\
h_i & = & \det(B\cdot A_1|\dots|B\cdot A_l|A_{l+1}|\dots|\widehat{A_i}|\dots|A_{n+1}) \\
(Type\ 3) & & j = 1, \dots, k \\
h_{n+1+j} & = & \det(B_j).
\end{array}$$

The degrees of the components are :

	degree
(Type 1)	$(k+1)\cdot(l-1)+1\cdot(n+1-l)$
(Type 2)	$(k+1)\cdot l+1\cdot(n-l)$
(Type 3)	n

As $h := \prod_{i=1}^{n+1+k} h_i$ is the reduced equation of the discriminant, we have

$$\begin{aligned}\deg(h) &= l\cdot\deg(Type\ 1)+(n+1-l)\cdot\deg(Type\ 2)+k\cdot\deg(Type\ 3)\\ &= nkl+n^2+nk+n.\end{aligned}$$

Hence the canonical equation of the discriminant is reduced if and only if

$$\begin{aligned} & \deg(h) &=& \dim(\mathrm{Rep}(Q,d))\\ \iff & nkl+n^2+nk+n &=& (n+1)n+kn^2\\ \iff & kl+n+k+1 &=& n+1+kn\\ \iff & kl+k &=& nk\\ \iff & n-1 &=& l.\end{aligned}$$

If $\#J=1$, then there is precisely one semi-invariant of $(Type\ 2)$. Namely,

$$h_{n+1} = \det(A_1|\ldots|A_n)$$

and the claim follows simply by counting the degrees again. □

We denote by $I_1=\{1,\ldots,n+1\}$ and $I_2=\emptyset$ and conclude directly from proposition 10.2.2 and proposition 10.2.3:

Corollary 10.2.4. The reduced discriminants of $\mathrm{Ext}_k\mathrm{Star}_n(I_i)$, $i=1,2$, are isomorphic and the same is true for the non reduced discriminants given by the canonical equations.

10.3 Calculations for $\mathrm{Ext}_k\mathrm{Star}_n$

In this section we present some results of our calculations of the spectrum and the Bernstein-Sato polynomial for $\mathrm{Ext}_k\mathrm{Star}_n(I)$ in specific cases. In order to compare them we consider only the canonical equation Δ of $\mathrm{Ext}_k\mathrm{Star}_n(I)$ here.

We briefly recall the notation for $\mathrm{Ext}_k\mathrm{Star}_n(I)$ from the previous sections:

n is the dimension of the vector spaces attached to the inner nodes

k equals the number of inner nodes or equivalently is equal to the number of maps B_i we inserted into the Star quiver $*_n$

$l = \#I$ is the number of sources we had attached to the sink on the left hand side,

i.e. we have the following picture in mind

$$1 \overset{A_I}{\rightrightarrows} n \overset{B_1}{\rightrightarrows} n \overset{B_2}{\rightrightarrows} \dots \overset{B_k}{\rightrightarrows} n \overset{A_J}{\leftleftarrows} 1.$$

In the following tables we list the Bernstein-Sato polynomials and the Spectrum at $t = 0$. Therefore we used the extended algorithm, as explained in the introduction of chapter 8.

l	k	n	**The Spectrum (at $t = 0$)**	Δ reduced
3	1	2	$(8, \frac{11}{3}, 2, 3, 4, 5, 6, 7, \frac{16}{3}, 1)$	no
2	1	2	$(\frac{10}{3}, \frac{13}{3}, 2, 3, 4, 5, 6, 7, \frac{14}{3}, \frac{17}{3})$	yes
1	1	2	$(\frac{10}{3}, \frac{13}{3}, 2, 3, 4, 5, 6, 7, \frac{14}{3}, \frac{17}{3})$	yes
0	1	2	$(8, \frac{11}{3}, 2, 3, 4, 5, 6, 7, \frac{16}{3}, 1)$	no
3	2	2	$(10, 11, \frac{16}{3}, 3, 4, 5, 6, 7, 8, 9, 10, \frac{23}{3}, 2, 3)$	no
2	2	2	$(\frac{14}{3}, \frac{17}{3}, \frac{20}{3}, 3, 4, 5, 6, 7, 8, 9, 10, \frac{19}{3}, \frac{22}{3}, \frac{25}{3})$	yes
1	2	2	$(\frac{14}{3}, \frac{17}{3}, \frac{20}{3}, 3, 4, 5, 6, 7, 8, 9, 10, \frac{19}{3}, \frac{22}{3}, \frac{25}{3})$	yes
0	2	2	$(10, 11, \frac{16}{3}, 3, 4, 5, 6, 7, 8, 9, 10, \frac{23}{3}, 2, 3)$	no
3	3	2	$(12, 13, 14, 7, 4, \dots, 13, 10, 3, 4, 5)$	no
2	3	2	$(6, 7, 8, 9, 4, \dots, 13, 8, 9, 10, 11)$	yes
1	3	2	$(6, 7, 8, 9, 4, \dots, 13, 8, 9, 10, 11)$	yes
0	3	2	$(12, 13, 14, 7, 4, \dots, 13, 10, 3, 4, 5)$	no
3	4	2	$(14, 15, 16, 17, \frac{26}{3}, 5, \dots, 16, \frac{37}{3}, 4, 5, 6, 7)$	no
2	4	2	$(\frac{22}{3}, \frac{25}{3}, \frac{28}{3}, \frac{31}{3}, \frac{34}{3}, 5, \dots, 16, \frac{29}{3}, \frac{32}{3}, \frac{35}{3}, \frac{38}{3}, \frac{41}{3})$	yes
1	4	2	$(\frac{22}{3}, \frac{25}{3}, \frac{28}{3}, \frac{31}{3}, \frac{34}{3}, 5, \dots, 16, \frac{29}{3}, \frac{32}{3}, \frac{35}{3}, \frac{38}{3}, \frac{41}{3})$	yes
0	4	2	$(14, 15, 16, 17, \frac{26}{3}, 5, \dots, 16, \frac{37}{3}, 4, 5, 6, 7)$	no

l	k	n	**The roots of the Bernstein-Sato polynomial $b_\Delta(s)$)**
3	1	2	$-\frac{9}{5}, -\frac{19}{15}, (-1)^6, -\frac{11}{15}, -\frac{1}{5}$
2	1	2	$(-\frac{4}{3})^2, (-1)^6, (-\frac{2}{3})^2$
1	1	2	$(-\frac{4}{3})^2, (-1)^6, (-\frac{2}{3})^2$
0	1	2	$-\frac{9}{5}, -\frac{19}{15}, (-1)^6, -\frac{11}{15}, -\frac{1}{5}$
3	2	2	$((-\frac{12}{7})^2, (-\frac{26}{21})^1, (-1)^8, (-\frac{16}{21})^1, (-\frac{2}{7})^2)$
2	2	2	$(-\frac{4}{3})^3, (-1)^8, (-\frac{2}{3})^3$
1	2	2	$(-\frac{4}{3})^3, (-1)^8, (-\frac{2}{3})^3$
0	2	2	$((-\frac{12}{7})^2, (-\frac{26}{21})^1, (-1)^8, (-\frac{16}{21})^1, (-\frac{2}{7})^2)$
3	3	2	$((-\frac{5}{3})^3, (-\frac{11}{9})^1, (-1)^{10}, (-\frac{7}{9})^1, (-\frac{1}{3})^3)$
2	3	2	$(-\frac{4}{3})^4, (-1)^{10}, (-\frac{2}{3})^4$
1	3	2	$(-\frac{4}{3})^4, (-1)^{10}, (-\frac{2}{3})^4$
0	3	2	$((-\frac{5}{3})^3, (-\frac{11}{9})^1, (-1)^{10}, (-\frac{7}{9})^1, (-\frac{1}{3})^3)$
3	4	2	$((-\frac{-18}{11})^4, (-\frac{40}{33})^1, (-1)^{12}, (-\frac{26}{33})^1, (-\frac{4}{11})^4)$
2	4	2	$((-\frac{4}{3})^5, (-1)^{12}(-\frac{2}{3})^5)$
1	4	2	$((-\frac{4}{3})^5, (-1)^{12}(-\frac{2}{3})^5)$
0	4	2	$((-\frac{-18}{11})^4, (-\frac{40}{33})^1, (-1)^{12}, (-\frac{26}{33})^1, (-\frac{4}{11})^4)$

l	k	n	**The Spectrum (at $t=0$)**	Δ reduced
4	1	3	$(15, 16, 11, \frac{21}{2}, \frac{31}{4}, \frac{35}{4}, 6, \ldots, 14, \frac{45}{4}, \frac{49}{4}, \frac{19}{2}, 9, 4, 5)$	no
3	1	3	$(\frac{21}{2}, \frac{23}{2}, \frac{29}{4}, \frac{33}{4}, \frac{37}{4}\frac{41}{4}, 6, \ldots, 14, \frac{39}{4}, \frac{43}{4}, \frac{47}{4}, \frac{51}{4}, \frac{17}{2}, \frac{19}{2})$	yes
2	1	3	$(\frac{21}{2}, \frac{47}{5}, 9, \frac{33}{4}, \frac{37}{4}, \frac{46}{5}, 6, \ldots, 14, \frac{54}{5}, \frac{43}{4}, \frac{47}{4}, 11, \frac{53}{5}, \frac{19}{2})$	yes
1	1	3	$(\frac{21}{2}, \frac{23}{2}, \frac{29}{4}, \frac{33}{4}, \frac{37}{4}\frac{41}{4}, 6, \ldots, 14, \frac{39}{4}, \frac{43}{4}, \frac{47}{4}, \frac{51}{4}, \frac{17}{2}, \frac{19}{2})$	no
0	1	3	$(15, 16, 11, \frac{21}{2}, \frac{31}{4}, \frac{35}{4}, 6, \ldots, 14, \frac{45}{4}, \frac{49}{4}, \frac{19}{2}, 9, 4, 5))$	no

l	k	n	**The roots of the Bernstein-Sato polynomial $b_\Delta(s)$)**
4	1	3	$(-\frac{12}{7})^2, (-\frac{10}{7})^1, (-\frac{19}{14})^1, (-\frac{33}{28})^2, (-1)^9, (-\frac{23}{28})^2, (-\frac{9}{14})^1, (-\frac{4}{7})^1, -(\frac{2}{7})^2$
3	1	3	$(-\frac{3}{2})^2, (-\frac{5}{4})^4, (-1)^9, (-\frac{3}{4})^4, (-\frac{1}{2})^2$
2	1	3	$(-\frac{3}{2})^1, (-\frac{7}{5})^1, (-\frac{4}{3})^1, (-\frac{5}{4})^2, (-\frac{6}{5})^1, (-1)^9, (-\frac{4}{5})^1, (-\frac{3}{4})^2, (-\frac{2}{3})^1, (-\frac{3}{5})^1, (-\frac{1}{2})^1$
1	1	3	$(-\frac{13}{7})^1, (-\frac{10}{7})^2, (-\frac{17}{14})^3, (-1)^9, (-\frac{11}{14})^3, (-\frac{4}{7})^2, (-\frac{1}{7})^1$
0	1	3	$(-\frac{12}{7})^2, (-\frac{10}{7})^1, (-\frac{19}{14})^1, (-\frac{33}{28})^2, (-1)^9, (-\frac{23}{28})^2, (-\frac{9}{14})^1, (-\frac{4}{7})^1, -(\frac{2}{7})^2$

In order to get the spectrum at $t \neq 0$ from the spectrum at $t = 0$ we apply Algorithm 2 (cf. proof of lemma 2.2.6 for the statement in terms of a base change or before remark 9.3.2 in terms of the spectrum at $t = 0$) and get

l	k	n	**The Spectrum (at $t \neq 0$)**	Δ reduced
3	1	2	$(\frac{8}{3}, 3, 2, 3, 4, 5, 6, 7, 6, \frac{19}{3})$	no
2	1	2	$(\frac{10}{3}, \frac{13}{3}, 2, 3, 4, 5, 6, 7, \frac{14}{3}, \frac{17}{3})$	yes
1	1	2	$(\frac{10}{3}, \frac{13}{3}, 2, 3, 4, 5, 6, 7, \frac{14}{3}, \frac{17}{3})$	yes
0	1	2	$(\frac{8}{3}, 3, 2, 3, 4, 5, 6, 7, 6, \frac{19}{3})$	no
3	2	2	$(4, 5, \frac{16}{3}, 3, 4, 5, 6, 7, 8, 9, 10, \frac{23}{3}, 8, 9)$	no
2	2	2	$(\frac{14}{3}, \frac{17}{3}, \frac{20}{3}, 3, 4, 5, 6, 7, 8, 9, 10, \frac{19}{3}, \frac{22}{3}, \frac{25}{3})$	yes
1	2	2	$(\frac{14}{3}, \frac{17}{3}, \frac{20}{3}, 3, 4, 5, 6, 7, 8, 9, 10, \frac{19}{3}, \frac{22}{3}, \frac{25}{3})$	yes
0	2	2	$(4, 5, \frac{16}{3}, 3, 4, 5, 6, 7, 8, 9, 10, \frac{23}{3}, 8, 9)$	no
3	3	2	$(6, 7, 8, 7, 4, \ldots, 13, 10, 9, 10, 11)$	no
2	3	2	$(6, 7, 8, 9, 4, \ldots, 13, 8, 9, 10, 11)$	yes
1	3	2	$(6, 7, 8, 9, 4, \ldots, 13, 8, 9, 10, 11)$	yes
0	3	2	$(6, 7, 8, 7, 4, \ldots, 13, 10, 9, 10, 11))$	no
3	4	2	$(8, 9, 10, 11, \frac{26}{3}, 5, \ldots, 16, \frac{37}{3}, 10, 11, 12, 13)$	no
2	4	2	$(\frac{22}{3}, \frac{25}{3}, \frac{28}{3}, \frac{31}{3}, \frac{34}{3}, 5, \ldots, 16, \frac{29}{3}, \frac{32}{3}, \frac{35}{3}, \frac{38}{3}, \frac{41}{3})$	yes
1	4	2	$(\frac{22}{3}, \frac{25}{3}, \frac{28}{3}, \frac{31}{3}, \frac{34}{3}, 5, \ldots, 16, \frac{29}{3}, \frac{32}{3}, \frac{35}{3}, \frac{38}{3}, \frac{41}{3})$	yes
0	4	2	$(8, 9, 10, 11, \frac{26}{3}, 5, \ldots, 16, \frac{37}{3}, 10, 11, 12, 13)$	no

l	k	n	**The Spectrum (at $t \neq 0$)**	Δ reduced
4	1	3	$(9, \frac{17}{2}, 8, 9, \frac{31}{4}, \frac{35}{4}, 6, \ldots, 14, \frac{45}{4}, \frac{49}{4}, 11, 12, \frac{23}{2}, 11)$	no
3	1	3	$(\frac{21}{2}, \frac{23}{2}, \frac{29}{4}, \frac{33}{4}, \frac{37}{4} \frac{41}{4}, 6, \ldots, 14, \frac{39}{4}, \frac{43}{4}, \frac{47}{4}, \frac{51}{4}, \frac{17}{2}, \frac{19}{2})$	yes
2	1	3	$(\frac{21}{2}, \frac{47}{5}, 9, \frac{33}{4}, \frac{37}{4}, \frac{46}{5}, 6, \ldots, 14, \frac{54}{5}, \frac{43}{4}, \frac{47}{4}, 11, \frac{53}{5}, \frac{19}{2})$	yes
1	1	3	$(9, 10, \frac{13}{2}, \frac{15}{2}, \frac{17}{2}, 8, 6, \ldots, 14, 12, \frac{23}{2}, \frac{25}{2}, \frac{27}{2}, 10, 11)$	no
0	1	3	$(9, \frac{17}{2}, 8, 9, \frac{31}{4}, \frac{35}{4}, 6, \ldots, 14, \frac{45}{4}, \frac{49}{4}, 11, 12, \frac{23}{2}, 11)$	no

Evaluation of $\mathrm{Ext}_k\mathrm{Star}_n(I)$

Let us summarize the calculations from above:
We note that the specturm of $\mathrm{Ext}_k\mathrm{Star}_n(I)$ consists of non negative rational numbers, which fulfil the symmetry

$$\nu_i + \nu_{n+1-i} = \deg(\Delta) - 1.$$

Moreover the spectrum at $t = 0$ and $t \neq 0$ coincides if and only if Δ is reduced. In contrast to the star quiver $*_n$, where the spectral numbers are always positive integers, this only holds in the above cases for $\mathrm{Ext}_3\mathrm{Star}_3$ (with any possible attachment of arrows to the sink on the left hand side).

The symmetry we observe for the serie $\mathrm{Ext}_k\mathrm{Star}_2$ is no surprise. Indeed we denote by $I_1 = \{1,2,3\}, I_2 = \{1,2\}, I_3 = \{1\}$ and $I_4 = \emptyset$, then we conlude from proposition 10.2.2 that the reduced discriminants of $\mathrm{Ext}_k\mathrm{Star}_n(I_1)$ and $\mathrm{Ext}_k\mathrm{Star}_n(I_4)$ (resp. $\mathrm{Ext}_k\mathrm{Star}_n(I_2)$ and $\mathrm{Ext}_k\mathrm{Star}_n(I_3)$) are isomorphic. Hence the same is true for any equation, if the multiplicities of the corresponding irreducible components are equal. In particular this is true for the canonical equation.

We believe that it is clear from the proof of proposition 9.4.1, which representation we choose[2] in each of the examples from above to check the conditions of the invariant subspace conjecture 8.1.1. We leave these easy calculations to the reader and conclude from the above tables:

Corollary 10.3.1. The serie $\mathrm{Ext}_k\mathrm{Star}_n(I)$ satisfies the invariant subspace conjecture 8.1.1 in each case from above.
The exponent condition 8.1.4 holds in the above cases if and only if the canonical equation Δ is reduced.

Because the star quiver $*_n$ and $\mathrm{Ext}_k\mathrm{Star}_n(I)$ are closely related, it is reasonable to adapt the ideas of chapter 9 in order to find formulas for the spectrum and the roots of the Bernstein-Sato polynomial for the serie $\mathrm{Ext}_k\mathrm{Star}_n(I)$.
In particular we hope to understand the effect on the spectrum and the roots of the Bernstein-Sato polynomial of the construction.
Unfortunately this is still work in progress. It seems that the crucial point is establishing a formula comparable to lemma 9.1.4 (or find a suitable basis of $\mathrm{Der}(-log\ h)$, when the canonical equation of $\mathrm{Ext}_k\mathrm{Star}_n)I)$ defines a linear free divisor). Then remaining proof should be similar to the one of the Star quiver, as we have a specific candidate for the linear function $f \in V^* \backslash D^*$.

[2]I.e. for the maps B_j we choose the identity.

Remark 10.3.2. At the moment calculations of the spectrum and the Bernstein-Sato polynomial of discriminants of $(\mathrm{Rep}(Q, \mathbf{d}))$ with big real Schur root $\mathbf{d}$ (e.g. the highest root of E_8) are out of reach due to memory limitations. But if we understand the effect of the construction, we should determine (at least some of) them. Indeed we can obtain many pairs $(E_8, \mathbf{d})$, where $\mathbf{d}$ is a positive sincere root of the Dynkin diagram E_8, by this construction from a pair $(E_7, \tilde{\mathbf{d}})$, where $\tilde{\mathbf{d}}$ is a suitable positive sincere root of E_7.

Bibliography

[ALMM10] Daniel Andres, Viktor Levandovskyy, and Jorge Martín-Morales, *Effective methods for the computation of Bernstein-Sato polynomials for hypersurfaces and affine varieties*, arXiv preprint arXiv:1002.3644 (2010).

[ARS95] Maurice Auslander, Idun Reiten, and Sverre O. Smalø, *Representation theory of Artin algebras.*, Cambridge: Cambridge University Press, 1995.

[Ben91] D.J. Benson, *Representations and cohomology. I: Basic representation theory of finite groups and associative algebras.*, Cambridge etc.: Cambridge University Press, 1991.

[Ber72] J.N. Bernstein, *The analytic continuation of generalized functions with respect to a parameter*, Functional Anal. Appl **6** (1972), no. 4, 26–40.

[Bjö79] Jan-Erik Björk, *Rings of differential operators*, North-Holland, 1979.

[Bjö93] ———, *Analytic D-modules and applications.*, Dordrecht: Kluwer Academic Publishers, 1993.

[BM06] Ragnar-Olaf Buchweitz and David Mond, *Linear free divisors and quiver representations*, Singularities and computer algebra **324** (2006), 41.

[BMS06] Nero Budur, Mircea Mustaţă, and Morihiko Saito, *Bernstein-Sato polynomials of arbitrary varieties.*, Compos. Math. **142** (2006), no. 3, 779–797.

[Bor87] Armand Borel, *Algebraic D-modules.*, Perspectives in Mathematics, Vol. 2, Boston etc.: Academic Press, Inc., 1987.

[Bou81] Nicolas Bourbaki, *Groupes et algèbres de Lie. Chapitres 4, 5 et 6.*, Elements de Mathematique. Paris etc.: Masson. 288 p., 1981.

[Bri12] Michel Brion, *Representations of quivers.*, Geometric methods in representation theory. I. Selected papers based on the presentations at the summer school, Grenoble, France, June 16 – July 4, 2008, Paris: Société Mathématique de France, 2012, pp. 103–144.

[CDS17] Alberto Castaño Domínguez and Christian Sevenheck, *Irregular Hodge filtration of some confluent hypergeometric systems*, preprint arXiv:1707.03259 [math.AG], 2017.

[CSS13] Sergio Caracciolo, Alan D. Sokal, and Andrea Sportiello, *Algebraic/combinatorial proofs of Cayley-type identities for derivatives of determinants and Pfaffians.*, Adv. Appl. Math. **50** (2013), no. 4, 474–594.

[Del70] Pierre Deligne, *Equations différentielles à points singuliers réguliers.*, vol. 163, Springer, Cham, 1970.

[Del73] _____, *Le formalisme des cycles évanescents*, Groupes de Monodromie en Géométrie Algébrique, Springer, 1973, pp. 82–115.

[DG70] Michel Demazure and Pierre Gabriel, *Groupes algébriques. Tome I: Géométrie algébrique. Généralités. Groupes commutatifs. Avec un appendice 'Corps de classes local' par Michiel Hazewinkel.*, Paris: Masson et Cie, Éditeur; Amsterdam: North-Holland Publishing Company. xxvi, 700 p., 1970.

[DG07] Ignacio De Gregorio, *Some examples of non-massive frobenius manifolds in singularity theory*, Journal of Geometry and Physics **57** (2007), no. 9, 1829–1841.

[DGPS18] Wolfram Decker, Gert-Martin Greuel, Gerhard Pfister, and Hans Schönemann, SINGULAR *4-1-1 — A computer algebra system for polynomial computations*, http://www.singular.uni-kl.de, 2018.

[DS03] Antoine Douai and Claude Sabbah, *Gauss-Manin systems, Brieskorn lattices and Frobenius structures. I.*, Ann. Inst. Fourier **53** (2003), no. 4, 1055–1116.

[DW00] Harm Derksen and Jerzy Weyman, *Semi-invariants of quivers and saturation for Littlewood-Richardson coefficients.*, J. Am. Math. Soc. **13** (2000), no. 3, 467–479.

[DZ01] M. Domokos and A.N. Zubkov, *Semi-invariants of quivers as determinants.*, Transform. Groups **6** (2001), no. 1, 9–24.

[Eis13] David Eisenbud, *Commutative algebra: with a view toward algebraic geometry*, vol. 150, Springer Science & Business Media, 2013.

[Gab72] Peter Gabriel, *Unzerlegbare Darstellungen. I. (Indecomposable representations. I).*, Manuscr. Math. **6** (1972), 71–103.

[GL91] Werner Geigle and Helmut Lenzing, *Perpendicular categories with applications to representations and sheaves.*, J. Algebra **144** (1991), no. 2, 273–343.

[GMNRS09] Michel Granger, David Mond, Alicia Nieto-Reyes, and Mathias Schulze, *Linear free divisors and the global logarithmic comparison theorem*, Ann. Inst. Fourier (Grenoble) **59** (2009), no. 2, 811–850.

[GMS09] Ignacio de Gregorio, David Mond, and Christian Sevenheck, *Linear free divisors and frobenius manifolds*, Compositio Mathematica **145** (2009), no. 05, 1305–1350.

[GMS11] Michel Granger, David Mond, and Mathias Schulze, *Free divisors in prehomogeneous vector spaces.*, Proc. Lond. Math. Soc. (3) **102** (2011), no. 5, 923–950.

[GS06] Michel Granger and Mathias Schulze, *On the formal structure of logarithmic vector fields*, Compositio Mathematica **142** (2006), no. 03, 765–778.

[GS10] ———, *On the symmetry of b-functions of linear free divisors.*, Publ. Res. Inst. Math. Sci. **46** (2010), no. 3, 479–506.

[Gyo91] Akihiko Gyoja, *Theory of prehomogeneous vector spaces without regularity condition.*, Publ. Res. Inst. Math. Sci. **27** (1991), no. 6, 861–922.

[Her03] Claus Hertling, *tt* geometry, Frobenius manifolds, their connections, and the construction for singularities.*, J. Reine Angew. Math. **555** (2003), 77–161.

[HHKU96] Dieter Happel, Silke Hartlieb, Otto Kerner, and Luise Unger, *On perpendicular categories of stones over quiver algebras.*, Comment. Math. Helv. **71** (1996), no. 3, 463–474.

[HT07] Ryoshi Hotta and Toshiyuki Tanisaki, *D-modules, perverse sheaves, and representation theory*, vol. 236, Springer Science & Business Media, 2007.

[Hum75] James E. Humphreys, *Linear algebraic groups.*, Graduate Texts in Mathematics. 21. New York - Heidelberg - Berlin: Springer-Verlag, 1975.

[Kac80] V.G. Kac, *Infinite root systems, representations of graphs and invariant theory.*, Invent. Math. **56** (1980), 57–92.

[Kac82] ______, *Infinite root systems, representations of graphs and invariant theory. II.*, J. Algebra **78** (1982), 141–162.

[Kas76] Masaki Kashiwara, *B-functions and holonomic systems*, Inventiones mathematicae **38** (1976), no. 1, 33–53.

[Kas83] ______, *Vanishing cycle sheaves and holonomic systems of differential equations*, Algebraic geometry, Proc. Jap.-Fr. Conf., Tokyo and Kyoto 1982, Lect. Notes Math. 1016, 134-142, 1983.

[Kat71] Nicholas M. Katz, *The regularity theorem in algebraic geometry.*, Actes Congr. internat. Math. 1970, 1, 437-443, 1971.

[Kat90] ______, *Exponential sums and differential equations*, Princeton, NJ: Princeton University Press, 1990.

[Kim03] Tatsuo Kimura, *Introduction to prehomogeneous vector spaces. Translated from the Japanese by Makoto Nagura and Tsuyoshi Niitani.*, Providence, RI: American Mathematical Society (AMS), 2003.

[KR86] H. Kraft and Ch. Riedtmann, *Geometry of representations of quivers.*, Representations of algebras, Proc. Symp., Durham/Engl. 1985, Lond. Math. Soc. Lect. Note Ser. 116, 109-145 , 1986.

[Lan02] Serge Lang, *Algebra. 3rd revised ed.*, 3rd revised ed. ed., New York, NY: Springer, 2002.

[Mal74] Bernard Malgrange, *On the polynomials of j. n. bernstein*, Russian Mathematical Surveys **29** (1974), no. 4, 81–88.

[Mal83] ______, *Polynômes de bernstein-sato et cohomologie évanescente*, Astérisque **101** (1983), 243–267.

[MM04] Philippe Maisonobe and Zoghman Mebkhout, *Le théorème de comparaison pour les cycles évanescents.*, Éléments de la théorie des systèmes différentiels géométriques, Paris: Société Mathématique de France, 2004, pp. 311–389.

[OV12] Arkadij L Onishchik and Ernest B Vinberg, *Lie groups and algebraic groups*, Springer Science & Business Media, 2012.

[Rin76] Claus Michael Ringel, *Representations of K-species and bimodules.*, J. Algebra **41** (1976), 269–302.

[RS17] Thomas Reichelt and Christian Sevenheck, *Non-affine Landau-Ginzburg models and intersection cohomology*, Ann. Sci. Éc. Norm. Supér. (4) **50** (2017), no. 3, 665–753.

[Sab87] Claude Sabbah, *Proximité évanescente. I: La structure polaire d'un $\mathcal{D}$- modules. (Vanishing neighborhood. I: The polar structure of $\mathcal{D}$- modules).*, Compos. Math. **62** (1987), 283–328.

[Sab97] ______, *Monodromy at infinity and fourier transform*, Publications of the Research Institute for Mathematical Sciences **33** (1997), no. 4, 643–685.

[Sab06] ______, *Hypergeometric periods for a tame polynomial.*, Port. Math. (N.S.) **63** (2006), no. 2, 173–226.

[Sab07] ______, *Isomonodromic deformations and Frobenius manifolds. An introduction. Transl. from the French.*, Berlin: Springer; Les Ulis: EDP Sciences, 2007.

[Sai80] Kyoji Saito, *Theory of logarithmic differential forms and logarithmic vector fields*, J. Fac. Sci. Univ. Tokyo Sect. IA Math. **27** (1980), no. 2, 265–291.

[Sai89] Morihiko Saito, *On the structure of brieskorn lattice*, Ann. Inst. Fourier (Grenoble) **39** (1989), no. 1, 27–72.

[Sat70] Mikio Sato, *Theory of prehomogeneous vector spaces*, Sugaku no ayumi **15** (1970)), no. 1, 85–156 (Japanese).

[Sch91] Aidan Schofield, *Semi-invariants of quivers*, Journal of the London Mathematical Society **2** (1991), no. 3, 385–395.

[Sev09] Christian Sevenheck, *Frobenius manifolds and variation of twistor structures in singularity theory*, https://www-user.tu-chemnitz.de/ sevc/Summary.pdf (2009).

[Sev11] ______, *Bernstein polynomials and spectral numbers for linear free divisors.*, Ann. Inst. Fourier **61** (2011), no. 1, 379–400.

[Sev13] ______, *Duality of Gauß-Manin systems associated to linear free divisors.*, Math. Z. **274** (2013), no. 1-2, 249–261.

[SK77] Mikio Sato and Tatsuo Kimura, *A classification of irreducible prehomogeneous vector spaces and their relative invariants*, Nagoya Mathematical Journal **65** (1977), 1–155.

[SS85] J. Scherk and J. H. M. Steenbrink, *On the mixed Hodge structure on the cohomology of the Milnor fibre*, Math. Ann. **271** (1985), no. 4, 641–665. MR 790119

[SSM90] Mikio Sato, Takuro Shintani, and Masakazu Muro, *Theory of prehomogeneous vector spaces. (Algebraic part). - The English translation of Sato's lecture from Shintani's note.*, Nagoya Math. J. **120** (1990), 1–34.